Overview of

Environmental Laws and Regulations

Navigating the Green Maze

Joseph J. Bernosky, P.E.

American Water Works Association

Overview of Environmental Laws and Regulations:
Navigating the Green Maze
First Edition

Author's Note: The opinions expressed in this book are the author's own and do not reflect the policies, practices, or opinions of the American Water Works Association, its members, other professional associations, or the author's employer, Board of Directors, or staff. Comments and criticisms should be addressed to the author alone.

AWWA Publications Manager: Gay Porter De Nileon
Project Manager/Technical Editor: Martha Ripley Gray
Cover Art: Jan R. Bailey and Daniel E. Feldman
Production: PerfecType, Nashville, TN

Library of Congress Cataloging-in-Publication Data
Bernosky, Joseph J.
Overview of environmental laws and regulations: navigating the green maze / Joseph J. Bernosky. — 1st ed.
p. cm.
Includes bibliographical references and index.
ISBN 978-1-58321-815-0
1. Water—Law and legislation—United States. 2. Sewage disposal—Law and legislation—United States. 3. Water—Pollution—Law and legislation—United States. 4. Environmental law—United States. I. Title.
KF5569.B47 2010
344.7304'6343—dc22

2010041177

Printed in the United States of America

American Water Works Association

6666 West Quincy Avenue
Denver, CO 80235-3098
303.794.7711
www.awwa.org

CONTENTS

PREFACE ix

About the Author x

1 INTRODUCTION 1

A Highly Regulated Society . . . 1
. . . And an Even More Highly Regulated Industry 3

2 THE GREEN MAZE 5

The Green Maze: What Is It? 5
Regulatory Structure 6
Not Getting Lost in the Green Maze 11

3 STAKEHOLDERS 13

Elected Officials 14
Regulators 16
The Utility 17

Customers (General Public) 17
Customers (Business, Industrial, Agricultural) 18
Media 18
Activists 19
A Special Stakeholder—the Environmental Protection Agency 20

4 A BRIEF HISTORY OF ENVIRONMENTAL REGULATIONS 23

Before the Modern Environmental Movement 26
The 1960s and 1970s 26

5 THE SAFE DRINKING WATER ACT 29

Preludes to the SDWA 31
The 1974 Safe Drinking Water Act 32

6 THE CLEAN WATER ACT 41

The National Pollutant Discharge Elimination System 43
The National Pretreatment Program 46
Water Quality Standards, the 303(d) Program, and TMDLs 47
Stormwater Regulations 48
Nonpoint Sources 51
Regulatory Intersection of the CWA and the SDWA 53

7 THE CLEAN AIR ACT 55

Risk Management Plans 59
Utilities and the Clean Air Act 59

8 HAZARDOUS WASTE AND HAZARDOUS MATERIALS REGULATIONS 61

History of Hazardous Waste Regulations and RCRA 63
Hazardous Materials Management and Right-to-Know Regulations 64
Water/Wastewater Utilities and Hazardous Waste 66

9 RECYCLED WATER 67

What Recycled Water Is 68
Recycled Water Uses 69
Federal Recycled Water Regulations 70
Indirect Potable Reuse 71
Graywater 73

10 OTHER LAWS AND REGULATIONS 75

The Endangered Species Act (a.k.a. Bugs and Bunnies) 76
Global Warming 79
Water Supply Availability Requirements 83
Bioterrorism Preparedness and Response Act 86

11 STATE ENVIRONMENTAL REGULATIONS 87

The Federal/State Regulatory Nexus 87
State Regulatory Structures 91

12 WATER LAW 95

Riparian Rights 96
Prior Appropriation (a.k.a. "First in Time, First in Right") 97
Interstate and International Water Compacts and Agreements 98

13 WATER AND WASTEWATER UTILITIES ORDINANCES 101

14 ENVIRONMENTAL REGULATIONS AND THE PUBLIC 105

Public Participation 105
Public Communication 107
Risk Communication 110
Environmental Advocacy 112

15 THE ROLE OF LABORATORIES 117

Laboratory and Custody Protocols 119
The Vanishing Zero 120

16 MORE MAZES: TOMORROW'S REGULATIONS 125

Emerging Contaminants 129

INTERNET RESOURCES 135

REFERENCES 137

ADDITIONAL READING 141

INDEX 143

PREFACE

This book originated with a class I taught as an adjunct instructor at Mount San Jacinto College, Calif. (http://www.msjc.edu/), during fall 2009. The class, WATR 130 Environmental Laws and Regulations, is an elective course within the college's Water Technology program. Water Technology is a certificate/Associate of Science program intended to prepare students for a variety of careers in water and wastewater operations. It's a great program and a great introduction to this important field.

And that's what this book is—an introduction. It's aimed at a variety of audiences: the fresh-out-of-college civil engineer who knows about reinforced concrete, the Hazen-Williams equation, and what a hydraulic grade line is, but who's never heard of the Safe Drinking Water Act. How about the bright young person who's just been promoted to a first-line supervisor in a wastewater utility's operations department? She knows the intricacies of the treatment plant, but not how the Clean Water Act is going to affect her professional life from now on. Or consider the seasoned legislative professional who knows the ins and outs of the political process, but very little about the regulatory world of water and wastewater utilities. This book's for him, too.

It's a simple (and perhaps slightly humorous) introduction to a very complex and very serious topic. It's meant to be a fairly easy read, and certainly is not intended to be an in-depth reference work for any of the myriad number of environmental laws and regulations discussed in these pages. There are other books and other courses for that level of detailed knowledge.

I'd like to thank three special people who had a great deal to do with my being a part of this industry: Dr. Jennifer Clancy, Dr. Michelle Frey, and Dr. Michael J. McGuire. Thanks for taking a chance; I wouldn't be here without your confidence and support.

Finally, for my wife, Nancy, thanks for putting up with the visits to pump stations, construction sites, half-completed storage tanks, and treatment plants—and thanks for always believing.

About the Author: Joseph J. Bernosky, P.E.

Joe Bernosky is director of engineering for Western Municipal Water District, a Metropolitan Water District of Southern California member agency. He has more than 25 years of professional experience with drinking water and wastewater utilities, industry, and consulting firms. During that time, he performed as a design engineer, project engineer/manager, and program manager. He has provided services to water utilities addressing water quality and distribution system issues, online water quality monitoring, water resources, residuals management, SDWA regulatory impacts, emergency response, and vulnerability assessments. He is a licensed professional engineer in California, Colorado, Oregon, Texas, and Washington.

CHAPTER 1

INTRODUCTION

A Highly Regulated Society . . .

We live in a highly regulated society. If you don't believe that statement, just check out this Web site, http://www.gpoaccess.gov/fr/index.html, on any weekday morning. That is the URL for the *Federal Register*, which is published every federal working day by the Office of the Federal Register, National Archives and Records Administration. The *Federal Register* (or *FR*) is the official publication for rules, proposed rules, and notices of federal agencies and organizations, as well as executive orders and other presidential documents.

Here's a sample from the *FR* (July 10, 2009):

Contents

Federal Register

Vol. 74, No. 131

Friday, July 10, 2009

Environmental Protection Agency

RULES

Approval and Promulgation of Air Quality Implementation Plans:

Texas; Revisions to the 1-Hour Ozone Plan for the Beaumont/Port Arthur Area, 33146–33152

Pesticide Tolerances:

Buprofezin, 33153–33159

Indoxacarb, 33159–33165

Mandipropamid, 33165–33170

PROPOSED RULES

Approval and Promulgation of Air Quality Implementation Plans:

California, 33196–33200

Texas; Revisions to the 1-Hour Ozone Plan for the Beaumont/Port Arthur Arca, etc., 33200

NOTICES

Environmental Impact Statements; Availability, etc., 33235–33237

This is the entry for the US Environmental Protection Agency (USEPA), just one of hundreds of federal agencies. Note the page numbers—33,235 for the last notice (one announcing the availability of an Environmental Impact Statement, or EIS); that means since January 1, 2009, more than 33,000 pages of proposed and final rules, regulations, notifications, and other federal actions had been published. By the end of 2009, the *FR* count reached nearly 70,000 pages (69,676 to be exact); a truly staggering total—and that's just the federal government. But 2009 still couldn't top 2008 when the *FR* contained 79,435 pages, an all-time record high. There are also numerous state and local regulations, especially in such highly regulated states as California and New Jersey.

. . . And an Even More Highly Regulated Industry

As regulated as American society is, some arenas are even *more* highly regulated—and this certainly includes the arena in which water and wastewater utilities operate. This book will attempt to explore that arena: that sometimes complex, sometimes confusing, but always challenging green maze of environmental laws and regulations. This book is not about the specific requirements of individual regulations or regulatory limits (such as the maximum contaminant level [MCL] for arsenic or nitrates in drinking water), or how to comply with those regulations. Instead, this book is intended to explore and describe that green maze of regulations and will address such topics as

- The green maze: What it is and how not to get lost
- The relationship of laws and regulations
- A brief history of environmental regulations
- Stakeholders, their roles, and relationships to each other
- The scope and nature of regulations impacting water and wastewater utilities
    - Safe Drinking Water Act
    - Clean Water Act
    - Clean Air Act
    - Hazardous waste regulations
    - Hazardous material regulations
    - Recycled water
    - Water law
    - Selected state environmental laws
    - Ordinances
- Risk communication
- Environmental advocacy and public participation
- Future regulatory trends

CHAPTER 2

THE GREEN MAZE

The Green Maze: What Is It?

Maze: (noun)

1. A confusing network of intercommunicating paths or passages; labyrinth.

2. Any complex system or arrangement that causes bewilderment, confusion, or perplexity.

3. A state of bewilderment or confusion.

4. A winding movement, as in dancing.

Origin: 1250–1300; ME mase, n. use of aph. var. of amasen to amaze

According to the preceding dictionary entry, a maze is a confusing network of intercommunicating paths or passages—a labyrinth. After working in the environmental field for some 25-odd years, I think the secondary definition—any complex system or arrangement that causes bewilderment, confusion, or perplexity—certainly seems more appropriate. I do know that meandering through that green maze certainly can leave one in a bewildered or confused state, which is the tertiary definition. I'm not sure I've ever seen an environmental professional dancing, so we'll just pass on the fourth definition.

Okay, so we know what a maze is, but what is a "green maze"? Well, I'll define it as the paths, passages, twists, and turns of federal, state, and local environmental laws—and the sometimes even more constraining and confusing paths, passages, twists, and turns of their associated regulations—the day-to-day directives, conditions, and required actions that ensure compliance with the underlying environmental law. As we'll see, the intent of these environmental laws was and remains noble: the protection of human health and the environment. But in practice, complying with the vast number of regulations associated with those noble laws has become a difficult, confusing, and certainly expensive proposition.

Maze theory asserts that if a maze is "simply connected" it can be solved by following the "left hand rule": upon entering the maze, just maintain contact with the left hand (or right hand) wall; you'll either find the exit or be led back to the original entrance. Well, unfortunately, the green maze is not simply connected. It is complex to the nth degree, and growing more complex every year.

The primary mission of water and wastewater utilities is public health—nothing we do is more important. As such, it is our industry's duty to not only comply, but go beyond compliance with environmental regulations intended to protect human health and the environment. But it's tough to understand and comply with those regulations when they appear to be nothing more than a confusing network of intercommunicating paths or passages that leave one in a state of bewilderment or confusion. So let's start solving the puzzle that is the green maze by trying to understand what the difference is between an environmental law and a regulation.

Regulatory Structure

What's an environmental law? What's an environmental regulation? What's the difference between the two—or are they the same thing? They're not, and the difference between the two is often misunderstood.

A legislative body (such as the federal congress or a state legislature) enacts laws. It is then up to either federal or state departments or administrative agencies to write regulations implementing the intent of the law (which itself is often a challenging task). Regulations are ancillary or subordinate to laws, but both are enforceable.

At the federal level, the US Code is the official compilation of codified laws, and the *Code of Federal Regulations* (*CFR*) is the official compilation of regulations. At a state level, such as in California, laws are codified in various codes (*California Administrative Code*, *California Water Code*, and so on) while the *California Code of Regulations* (CCR) is the official compilation and publication of the regulations adopted, amended, or repealed by California state agencies. Other states have similar arrangements. For example, in Illinois, environmental regulations for the state are found in Title 35 of the *Illinois Administrative Code*; in New Jersey, it's Title 7 of the *New Jersey Administrative Code*, and so on.

The federal *CFR* is compiled into "Titles," which cover such diverse areas as agriculture, health and safety, the environment, and so forth. Here's a link to the *CFR*:

- *Code of Federal Regulations:*
  http://www.gpoaccess.gov/cfr/index.html

Stop reading, go over to your computer, open a web browser, and type in the above URL. Look around, get to know the various titles and what they cover, but be sure and check out Title 40—Protection of Environment; it's important. Go ahead. I'll wait . . . . Complicated, and just a bit daunting, eh?

Online and immediate access to the *CFR* is a boon to environmental professionals. Once upon a time (and I know for you younger folks this is a difficult concept to grasp) before the Internet, you had to purchase the individual titles on a yearly basis—and you still can at http://bookstore.gpo.gov/baskets/cfr-listing.jsp. The colors on the spine change every year (or at least they used to) and the *CFR* section in many consultants' libraries looked like a badly composed rainbow with spines in red, aquamarine, or brown.

From Laws to Regulations

Otto von Bismarck once famously said, "Laws are like sausages, it is better not to see them being made." The roundabout journey from a legislator-introduced bill to an adopted law can be a messy process with various interest groups (citizens, public interest groups, businesses, or other government agencies—all stakeholders, more on them in the next chapter) participating in the process. These stakeholders are often represented by lobbyists, individuals who are paid to explain their stakeholders' particular concerns or interests in the legislation. Because of these competing interests, a bill that is relatively simple in theory can become quite complex and complicated by the time it is finally passed into law.

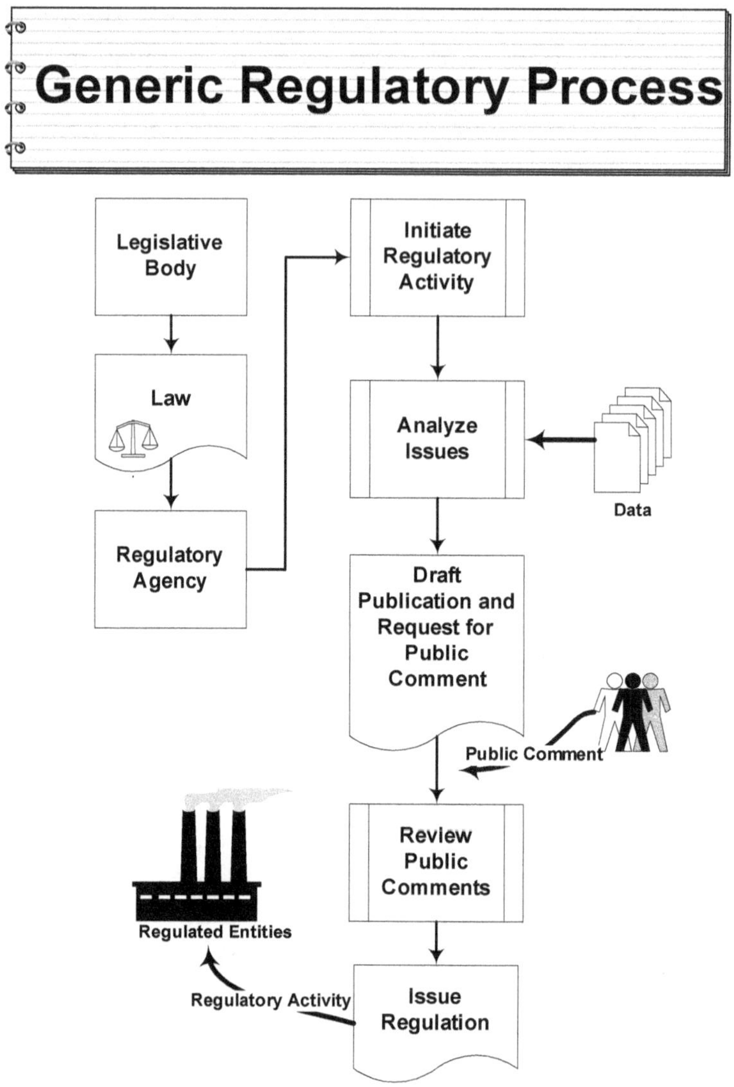

Once it becomes law, the next step is turning the statutory language into regulatory language—in other words what a regulated entity must do to comply with the law. This process of determining the language and terms of a law and putting it into action is known as *promulgation*.

The following is a generalized sequence of regulatory promulgation activities that are followed (roughly) at both the federal and state levels:

1. ***Initiate Regulatory Activity***—Operating under the statutory authority (the law itself, such as the Clean Water Act), an agency commences regulatory activity.

2. ***Analyze the Issues***—The agency evaluates the law and associated issues. Staff outlines the major questions that must be answered, the data needed, the experts who should be consulted, the anticipated costs, and other rulemaking needs. The agency may assemble a panel of industry or academic experts, or request input from the regulated community.

3. ***Draft Publication and Request for Public Comment***—At the federal level, a draft publication is issued that requests comments from the regulated community. A copy of the proposed regulation is published (in the *FR*) and a public comment period (typically 60–90 days) initiated. A similar process is followed at the state level with comment periods of varying lengths.

4. ***Review Public Comments***—The agency reviews and evaluates the received public comments. Depending on the regulation, these comments may range from recommendations for minor clarifications or corrections to extensive rewriting. The agency analyzes and may incorporate comments or changes before the draft final regulation is developed. Depending on the complexity (and contentiousness) of the regulation, this process could take years. On some occasions, there have been substantial and substantive differences between the proposed and final rules.

5. ***Issue Regulation***—The regulation is formally promulgated and published. Effective dates vary. A regulation may be effective on the day it is published or it may be effective a year later. These dates are specified in every regulation.

Environmental professionals spend the majority of their time wandering through the maze of regulations as opposed to the underlying laws themselves. Typically, that purview lies with environmental attorneys. Regulations detail the how and why of compliance with the law; only rarely will there be a need to delve into the specific details of the environmental law itself—usually (and unfortunately) when noncompliance results in litigation.

Permits

So we have an environmental law and now a regulation (or set of regulations); what next? What's next is one mechanism for enforcing a regulation—particularly one in which there are quantitative limits on discharges to water or air emissions. That mechanism is known as an environmental permit, which is a legal document required by a regulation and issued by the governing regulatory agency that allows the regulated entity (or permittee) to emit or discharge a pollutant or engage in certain regulated activities such as providing drinking water or operating a wastewater treatment plant. Pretty straightforward, right? Well, sometimes yes, sometimes no.

Most of the regulations that we'll review involve some type of permit—air permits, wastewater or stormwater discharge permits, drinking water permits. All of these are required for the permittee to operate a boiler, a wastewater treatment plant, or a drinking water utility. So what's in a permit?

Depending on the regulation and the regulatory agency, a permit can either be quite simple or extraordinarily complex. Permits essentially detail the requirements for operating under the applicable regulation. They may specify certain operating hours (some air permits), how pollution controls are expected to operate (again, air permits), required analytical testing protocols and frequencies (most permits including drinking water), and, probably most importantly, discharge limits (most air and water-related permits).

Much like the nature of the environmental permit itself, obtaining one through what is known as the permitting process can either be quite simple or extraordinarily complex. Some permit applications require a simple form to be filled out with administrative information and identifying the process, process controls, and so forth. If the permitted device or operation meets statutory requirements—bam!—you get the permit. In other cases, the permitting process can be quite protracted, requiring the collection and presentation of detailed data during lengthy negotiations with the regulatory agency. In some cases, the permitting process can literally take years.

Policy

One definition of *policy* is *a deliberate plan of action to guide decisions and achieve rational outcomes.* Not every instance or circumstance can (or should) be addressed by the regulations. This is where the concept of a policy comes into play. Often, the administering agency issues policy statements that outline its approach to an issue or interprets some aspect of regulatory language. Regulatory agencies must be cautious in their use of policies. Occasionally (and unfortunately), these policies extend beyond the formally promulgated regulatory structure and become "underground law."

Regulatory Costs

It is inarguable that the past 35 years or so of modern environmental regulations have provided many benefits—cleaner air, cleaner water, safer drinking water, and general environmental protection—all resulting in a higher quality of life for Americans. But there are related costs as well. As we'll discover, complying with regulations doesn't just mean purchasing and installing some type of pollution control equipment; the programmatic costs of compliance can be enormous. What are programmatic costs? They're the testing, inspecting, reporting, and certifying processes that demonstrate ongoing compliance with environmental regulations. And that's just half of the compliance cost story: local, state, and federal agencies also spend billions to ensure regulatory compliance.

The Competitive Enterprise Institute (Crews 2009) cites a report prepared for the Small Business Administration, Office of Advocacy, by economists W. Mark Crain and Thomas D. Hopkins that estimates 2008 federal regulatory compliance costs at $1.172 trillion (keep in mind that's *all* federal regulations—food, health, safety, EEOC issues, automobiles, and so on—not just environmental regulations). That mammoth sum is 8 percent of the US gross domestic product (GDP). Is that a bargain? Is the American public receiving benefits equal to $1.172 trillion? A *cost–benefit analysis* attempts to answer that question: do the estimated costs of a proposed regulation equal its projected benefits?

Although some folks believe differently, even public health and environmental regulations are subject to cost–benefit analyses. That means agencies evaluate the costs of regulating *something* at a *certain level* and try to determine if the benefits equal (or exceed) those costs. Sometimes, it's those last tiny, incremental benefits that cost the most. Again, are they worth it? That's a difficult question when an agency is trying to determine if the appropriate risk is 1×10^{-5} or 1×10^{-6}, and I guess the real issue is whether you happen to be that "1." (We'll discuss risk assessment and communications in a later chapter.) But keep this concept firmly in mind—the need to evaluate the costs and benefits of environmental regulations—as we proceed to explore various corridors in the green maze.

Not Getting Lost in the Green Maze

OK, so we're in this green maze of environmental laws and regulations; and we've already acknowledged that it's not "simply connected," so we can't just follow the left hand rule. How do we avoid getting lost? More importantly, how do we get from "enter" to "exit"? Well, the first thing you

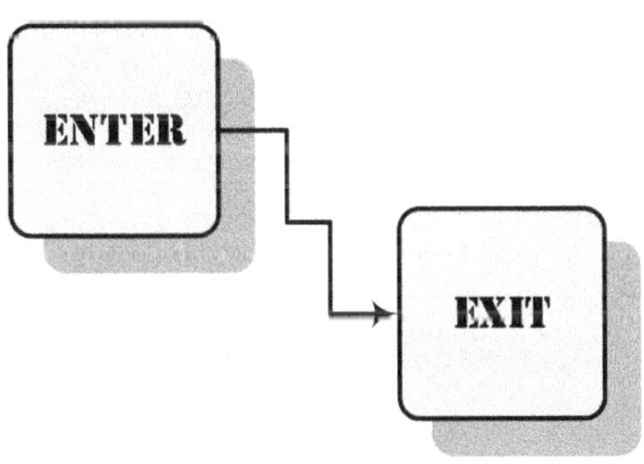

should know is—you can't get out. If you are in, and plan on remaining in, the water or wastewater field, you'll be in this green maze for the rest of your career. The best you can hope for is not getting totally lost.

Getting to know the other folks wandering through this maze and why they're in there can't hurt. Also, a general understanding of environmental laws and regulations—where and why they originated, their intent, and their scope—won't hurt either. And hopefully that's what the balance of this book will accomplish: helping a water or wastewater industry professional navigate the green maze.

CHAPTER 3

STAKEHOLDERS

Before embarking on our Theseus-like journey through the green maze, let's take a look at who we might encounter around the next corner. These are your fellow wanderers in the green maze. We'll call them *stakeholders*. Who are they? Why are they in the maze? What are their desired outcomes, their characteristics, their biases?

There are many definitions of a stakeholder:

- An individual, group, or organization impacted by the decisions and actions of an organization.
- A person with a vested interest in an organization or the successful completion of a project.
- "Any individual or organization that has an interest in water management activities. In the broadest sense, everyone is a stakeholder, because water sustains life. Water resources stakeholders are typically those involved in protecting, supplying, or using water for

any purpose, including environmental uses, who have a vested interest in a water-related decision" (Utah Division of Water Resources 2010).

For the purposes of this book, we will define the following as stakeholders in the green maze:

- Elected officials
- Regulators
- Utility (project or program proponent)
- Customers (general public)
- Customers (business, industrial, agricultural)
- Press/media
- Activists

Theseus with the Head of the Minotaur.
Photo courtesy of Plutho

The graphic on page 15 illustrates some basic relationships among the various stakeholders. As you can see, it's pretty complex, and these relations are not static. They strengthen, weaken, break, re-form, and change as circumstances and stakeholders change.

As we explore the green maze of assorted environmental regulations, policies, and issues, keep in mind the characteristics of the various stakeholders, their motivators and drivers. And remember what Shakespeare wrote (*King Henry VIII*, Act 3, Scene 2), "'Tis well said again; And 'tis a kind of good deed to say well: And yet words are no deeds."

Before we take a look at the stakeholders, remember that assuming broad generalizations about the characteristics of *any* group can be treacherous (and often wrong). But if we're to understand their individual roles in legislating, developing, promulgating, enforcing, or complying with environmental laws and regulations, we need to establish some sense of who these stakeholders are and what's driving their actions.

Elected Officials

For our purposes, elected officials are considered to be federal and state legislators, city councils, county supervisors, governors, mayors, and utility board members. All must stand for periodic election and therefore are answerable to voters—the public. What are the characteristics of an elected

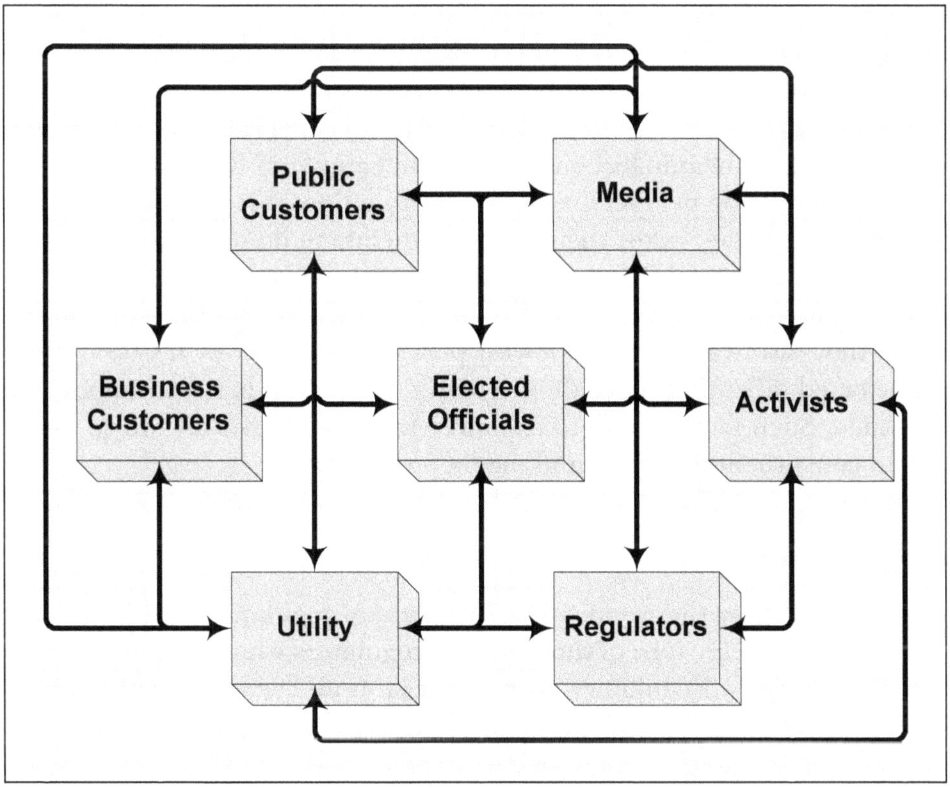

official? Well (and remember the admonition about generalizations), some common characteristics might be:

- Constituent sensitive
- Media sensitive
- Support sensitive
- Desire to avoid controversy

Obviously, this is a limited set of descriptors and there could be many more.

So, what are their motivators and drivers? Are they driven by a sense of duty or a devotion to public service? For a very few, could there be other less noble motivators in play such as ego, power, recognition? Does a particular elected official have a particular agenda, or a pet issue or cause? Because of constituent sensitivity, there is often a reluctance to take on an unpopular or a hot issue.

At the national level, politicians are currently held in pretty low repute by the general public. A September 2010 Gallup poll (Gallup 2010) reported that Congress's approval rating was a paltry 20 percent (with 77 percent expressing disapproval). The accompanying story noted that "Americans show no signs of relenting in their broad disapproval of the job Congress is doing as the midterm elections draw near." A poll conducted by the same organization just one year earlier revealed a similar level of discontent, attributing the low ratings to the possibility "that Americans are frustrated with the lack of meaningful outcomes from Congress" (Gallup 2009).

So there's one perspective on the bottom line; this poll suggests that voters—the public—want their elected officials to do *something*—or perhaps in the case of local water utility boards, *not do something* (as in raise rates). But the public is fickle; while they're demanding that *something*, they also want it at the lowest possible (or no) cost. It's no wonder that elected officials believe they're being whipsawed between impossible demands. Such factors help to explain why elected officials are so sensitive to constituent demands and media pressures.

Regulators

There are literally thousands of environmental regulations in the United States, and there are tens of thousands of regulators who write and enforce those regulations. Regulatory agencies exist at all levels of government—federal, state, and local. In many (in fact, most) cases, it was always the intent for federal regulations such as the Clean Water Act and Clean Air Act to be enforced at the state level.

Just as with elected officials, there is no such thing as a "typical" regulator, but some common characteristics might be:

- Technical
- Risk averse
- Usually detail oriented

Depending on the agency, regulators may report to appointed officials (such as California's Air Quality Management Districts' Boards) and as such a political component exists there as well. Motivators and drivers vary: a dedication to the environment, altruism, scientific inquiry, or a desire to improve their communities.

Many folks in the environmental field believe that environmental regulators have a great deal of authority and power, and that's true to some extent. However, such authority and power are constrained by the scope of the regulations, the authority delegated by the legislature (the underlying environmental law itself), and the courts' interpretations of those laws and regulations.

One of the most important relationships in the green maze is that between the regulator and the utility (or other regulated entity). The regulators who oversee water and wastewater utilities are their partners in public health. Thus it's important that utility staff invest considerable effort and resources to develop a relationship based on mutual respect and, more importantly, on trust.

The Utility

A water or wastewater utility is more of a thing than a person, but as with any organization, it is composed of people who have their individual characteristics. And, at least for public utilities, there is a political element as well. But what about the utility itself; what are some characteristics of a typical water or wastewater utility?

- Risk averse
- Customer service oriented
- Strong sense of mission and public service

What are motivators and drivers for the typical utility? Paramount must be regulations, principally the Safe Drinking Water Act and/or the Clean Water Act. But bear in mind that there are both public water and wastewater utilities (special districts such as Eastern Municipal Water District in Perris, Calif., or city-owned such as Austin Water [Austin, Texas]) and investor-owned utilities (IOUs), so that there may be a profit motive as well.

I have worked in a number of industrial sectors but have never met staff more dedicated than those who provide safe drinking water and efficient wastewater services. More than any other trait, this dedication to public health, I believe, is the primary driver for water and wastewater utilities.

Customers (General Public)

One class of water or wastewater utility customers is the general public. There are millions of customers and it might be a ridiculous question to ask, but what might be some broad-brush characteristics of the general public as utility customers?

- Probably not technically savvy
- Low risk threshold (in some cases approaching zero!)
- Seemingly contradictory desires (i.e., highest quality water at lowest possible price)
- Sense of entitlement?
- Demand for almost immediate gratification?

What are some motivators and drivers for the general public? For most customers it's probably pretty straightforward: they want safe, reliable, and inexpensive water and wastewater services. However, into that relatively simple mix, you have to add some complicating characteristics as noted previously.

Most of the general public doesn't know or care about disinfection by-products; all they care about is safe water—you can skip the details. Yet sometimes, a utility is *required* to get into the details, such as with the annual Consumer Confidence Report (or CCR), which is discussed later in the book. Other times, attending to those details requires considerable outlay of capital—and that means rate increases.

Thus we're dealing with a potentially difficult stakeholder here with the general public: fickle, technically unsophisticated, with conflicting desires; all of these traits jumbled together mean that at times the relationship between customers and the utility (or elected officials) could be tenuous. But much like a utility's relationship with its regulators, it's imperative that utilities maintain a strong sense of trust with their customers. Anything that breaches that trust has the potential to derail the utility's mission of public health.

Customers (Business, Industrial, Agricultural)

Why should we differentiate this class of utility customers from the general public? Don't they share similar characteristics? Well, yes and no. You'd be hard pressed to find a business customer opting for higher rates. But as business owners and operators, they have a better grasp of finances and understand the need for rate increases. So although they may share some characteristics, others differ:

- Probably more technically savvy than the general public
- Better understanding of risk and cost–benefit analyses
- Although desirous of low-cost commodities (water/wastewater services), reliability and predictability of rates (and thus their costs) are more important

There are also some differences between institutional customers; for example, in many areas, agricultural water has historically been priced below market value. In many states, agricultural interests (and lower water costs) may compete with and oppose industrial interests and residential customers.

Media

Mass media (specifically news organizations) in today's world are in a state of flux. The traditional news media of print newspaper and television are especially hard pressed to compete with what is becoming known as the new media. A generation ago, the word *Internet* didn't exist; today it may represent an entirely new and original news paradigm. For the purposes of this discussion, we'll limit our review to the more traditional media: television/radio news and print media. But with the advent of the 24-hour news cycle, even television news has changed over the past 10–15 years.

However, our discussion shouldn't completely overlook the new media, especially the "citizen journalist." With today's pocket technology, anyone with a cellular telephone camera and access to the Internet can become a citizen journalist and break real news.

As for the more traditional media, what are some general characteristics?

- "If it bleeds, it leads" (controversy sells)
- Competitive
- Often deadline driven (again especially with the 24-hour news cycle)
- News in the public interest

What motivates the press? Although some press coverage may be agenda driven, there is the need to sell papers or earn higher ratings. Just as individuals have many inherent agendas and biases, there are many facets of agenda journalism—altruism, public interest, desire for the truth, and political allies/advancement are just some of those facets.

Activists

There is an entire spectrum of environmental activists, all shades and colors and hues (see http://en.wikipedia.org/wiki/Environmentalism). But again, we'll need to limit our examination to those activists who populate just the particular maze encompassing water/wastewater utilities. What are they interested in or concerned with? Rates? Chemical constituents in water? Using drinking water reservoirs for public recreation? As their causes vary, so their characteristics may vary as well, but some common traits may be:

- Rational beings, who may occasionally engage in (seemingly) irrational acts
- Highly motivated (to their particular cause or issue)
- Dedicated (to their particular cause or issue)
- Bright and resourceful
- Often taking an issue personally

Their motivators and drivers are fairly straightforward: they want to achieve their (often singular) goal. Depending on the nature of their issue, it may be difficult to communicate with activists on a reasoned basis—in other words, facts (especially science- or risk-based facts) are completely discounted. Such an attitude may make conflict resolution difficult. Because the activist's issue is often personal, appealing to the greater good of the community may not be an effective negotiating strategy.

A Special Stakeholder: The Environmental Protection Agency

Perhaps the biggest and most well-known stakeholder running through the green maze is the federal Environmental Protection Agency (USEPA). And it is for this reason that we'll briefly examine the primary federal agency responsible for environmental protection and thus environmental regulations.

As will be discussed in the next chapter, prior to the creation of the USEPA, a number of federal agencies were responsible for regulating water pollution and air pollution, most notably the US Public Health Service (which is now under the auspices of the US Department of Health and Human Services and performs excellent work in the field of public health). In summer 1970 under Reorganization Plan No. 3 of 1970, the Nixon Administration announced plans to establish the USEPA (and incidentally, the National Oceanic and Atmospheric Administration as well). The USEPA itself was created on December 4, 1970, with William Ruckelshaus named its first administrator.

Thousands of federal employees and their responsibilities were transferred from existing federal agencies to the new environmental big dog, among them:

- Department of the Interior
  - Federal Water Quality Administration
  - Functions vested in the Secretary or Department by the Federal Water Pollution Control Act
  - Water Pollution Control Advisory Board, and the hearing boards provided for in the Federal Water Pollution Control Act
- Department of Agriculture
  - Functions of the Secretary or Department under the Federal Insecticide, Fungicide, and Rodenticide Act
- Department of Health, Education and Welfare

- ▲ Functions vested by law in the Secretary or Department administered through the Environmental Health Service, including the National Air Pollution Control Administration, Bureau of Solid Waste Management, Bureau of Water Hygiene, and Bureau of Radiological Health
- The Air Quality Advisory Board
- Atomic Energy Commission
  - ▲ Division of Radiation Protection Standards
- Council on Environmental Quality (CEQ)
  - ▲ Functions of CEQ under the National Environmental Policy Act (NEPA) of 1969

Ruckelshaus didn't sit back but came out swinging almost immediately, notifying both private corporations and public agencies that they would be held responsible for their environmental misdeeds. For example, he warned the cities of Atlanta, Detroit, and Cleveland to halt alleged violations of water pollution standards within 180 days or face possible federal suits. Atlanta was cited for what he termed "the massive discharge of pollutants"

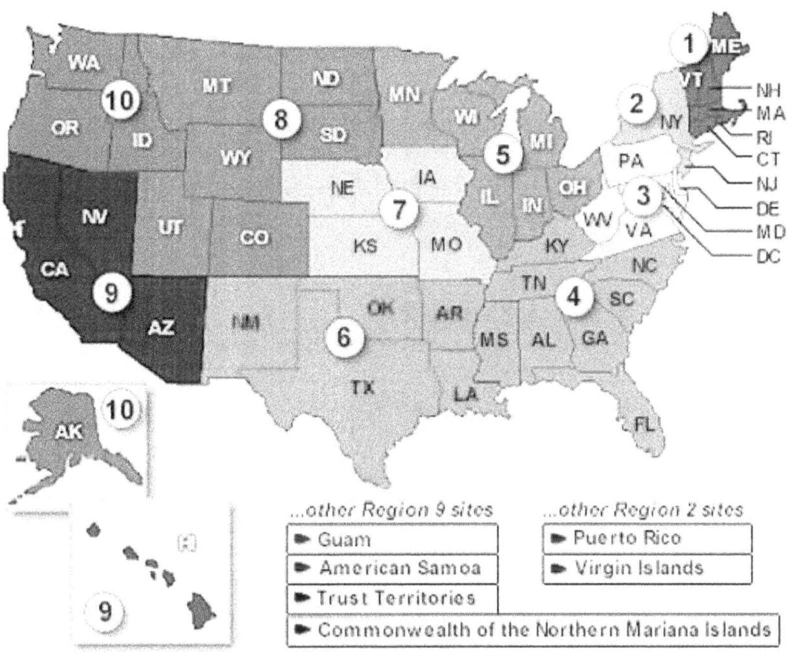

Courtesy of USEPA

into the Chattahoochee River (Wisman 1985); Detroit and Cleveland for polluting Lake Erie. Since then, the agency has never looked back.

Headquartered in Washington, D.C., the 18,000 employees of the USEPA carry out their duties through 10 regional offices. As the diagram on p. 21 illustrates, individual offices are responsible for both internal administrative tasks and various environmental media (air, water, and so on).

Functional Offices—USEPA

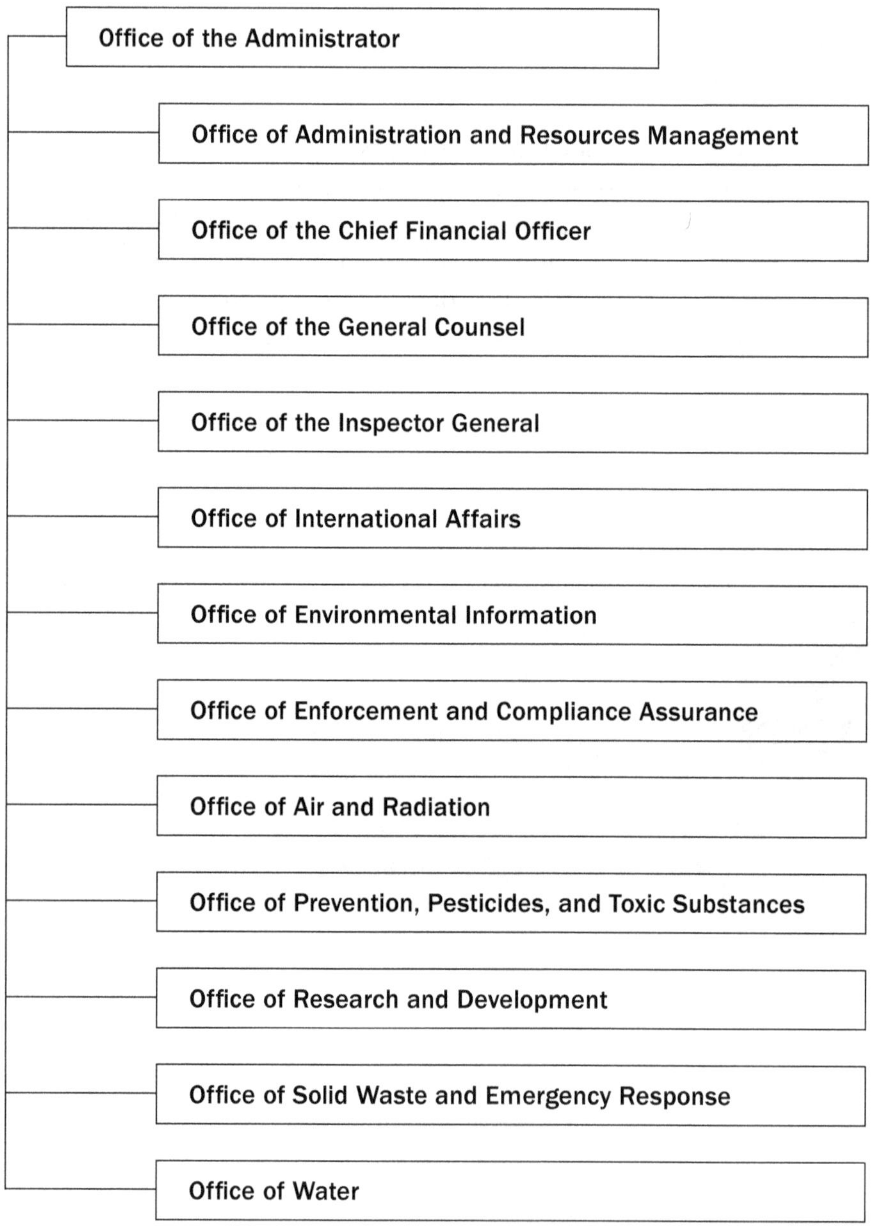

CHAPTER 4

A BRIEF HISTORY OF ENVIRONMENTAL REGULATIONS

There is a common fallacy that the modern age of environmental regulations began only in the 1960s and early 1970s, emerging from Washington, D.C., as Athena sprang fully formed from the head of Zeus.

I believe a stronger case can be made that the development of environmental regulations parallel 1) a society's wealth, 2) the expanded use of technology in that society, and 3) the increasing density of urban population centers. Ancient civilizations recognized the potential harm of turbid (dirty) water or smoke-filled air, but regulations as such were not enacted. The timeline on pages 24–25 illustrates the parallel developments of technology and regulations (or at least royal orders!):

Late Middle Ages	1300	1306: Edward I prohibits use of sea coal while Parliament is in session.
		1661: Antismoke tract *Fumifugium*
	1800	
Industrial Revolution		
First modern water and wastewater systems		1875: England, a Public Health Act contains a smoke abatement section.
Growth of the chemical industry	1900	1899: Refuse Act of 1899 (River and Harbor Act of 1899) prohibits discharges into navigable rivers.
Early Twentieth Century Foundation of today's modern chemical industry		1912: Public Health Service Act investigates relationship of water pollution to human diseases.
First use of chlorine for disinfection in United States—1908		1914: US Public Health Services Standards for biological contamination in drinking water. Revised 1925, 1946, and 1962.
World War II		1924: Oil Pollution Control Act
Large-scale chemical synthesis		1948: Water Pollution Control Act, first major water/ wastewater law, funding for infrastructure, Dept. of Justice suit. Amendments in 1952 and 1956.
Pesticides		
Automobiles		
Nuclear technology	1960	1955: Air Pollution Control Act. Extensions in 1959.
Waste disposal issues		1961: Federal Water Pollution Control Act

1960s to Date

	1960	1963: Original Clean Air Act, granted permission for federal air pollution control activities
Beginnings of modern environmental movement		
Semiconductors		
Clean industries		1969: National Environmental Policy Act
Santa Barbara oil spill		1970: US Environmental Protection Agency established
Cuyahoga River	1970	
		1970: Amended Clean Air Act (basis for modern CAA)
Increasing analytical capabilities		1972: Water Pollution Control Act (modern Clean Water Act)
	1975	1974: Safe Drinking Water Act, amended in 1986 and 1996
Love Canal		1976: Resource Conservation and Recovery Act (hazardous waste management regulations)
Personal computers		1980: Comprehensive Environmental Response, Compensation and Liability Act (CERCLA or Superfund)
Bhopal, India		Late 1980s: Various hazardous material management and right-to-know regulations
Part-per-trillion analytical capabilities	2000–	2000+: Greenhouse gas emission controls?
The Internet		

Before the Modern Environmental Movement

As the preceding chart illustrates, "environmental regulations" (of a sort) were enacted well before the 1960s and 1970s; and those listed are just a small sample of early regulations. There also were many individuals—known as naturalists or conservationists—who pioneered a love and appreciation of nature and the environment that eventually resulted in the awareness that led to truly protective environmental regulations:

- Alexander Wilson (1766–1813)—Born in Scotland, an early American poet and ornithologist who predicted some bird extinctions
- James J. Audubon (1785–1851)—Perhaps America's most famous ornithologist and illustrator (original copies of his seminal large-format folio *Birds of America* are worth millions of dollars)
- John Muir (1838–1914)—Also Scottish-born, an American naturalist, author, and early advocate of preserving US wilderness; founded the Sierra Club
- John Burroughs (1837–1931)—An American literary naturalist and essayist known at the turn of the twentieth century as the Grand Old Man of Nature
- Gifford Pinchot (1865–1946)—Reformer of the management and development of US forests, and advocate for the conservation of the nation's forest reserves by planned use and renewal; first chief of the US Forest Service

The 1960s and 1970s

One of the most famous icons of the 1960s environmental movement was Rachel Carson, who wrote *Silent Spring*. She is often cited as one of the founders of the modern environmental movement. In her 1962 book, Carson sounded the warning that widespread use of DDT (dichloro-diphenyl-trichloroethane, a pesticide first compounded in 1874) was harming songbirds and raptors. She was a gifted writer, and told a compelling tale. Her research and message were attacked by a number of industry groups, but the use of DDT in the United States was nonetheless banned in 1972. The other side of this environmental tale is that malaria, whose vectors were once well controlled by DDT, continues to be a deathly scourge around the world.

Another icon of the modern environmental movement was the 1969 Cuyahoga River fire, which many believe led to the creation of the USEPA and eventually the modern CWA. But it seems at the time, the fire was not regarded as that big a deal in Cleveland. A fortieth-anniversary story on the fire related that "the story barely made the newspapers the next day." It was quickly brought under control by crews on a firefighting tugboat. The river had burned before, as early as 1868, and several times more before a major fire in 1952 caused extensive damage to docks and nearby structures (Scott 2009). But whatever the cause or extent of the fire on the Cuyahoga River in June 1969, it ignited flames that probably did contribute to passage of the CWA some three years later.

Indeed, before the passage of the 1972 CWA and the massive infusion of federal funds for constructing new wastewater treatment plants, many municipalities routinely discharged only primary-treated sewage and other detritus into the nation's waterways.

But there's no question that the 1960s and early 1970s did usher in the era of modern environmental laws and regulations. Although some histories point to a distinct harbinger, such as *Silent Spring*, the Cuyahoga River fire, or incidents such as the 1969 Santa Barbara oil spill, in truth there was no one single, underlying causal factor driving environmental concern. Rather it was a number of disparate factors that combined to generate an environmental zeitgeist that resulted in so many seminal regulations.

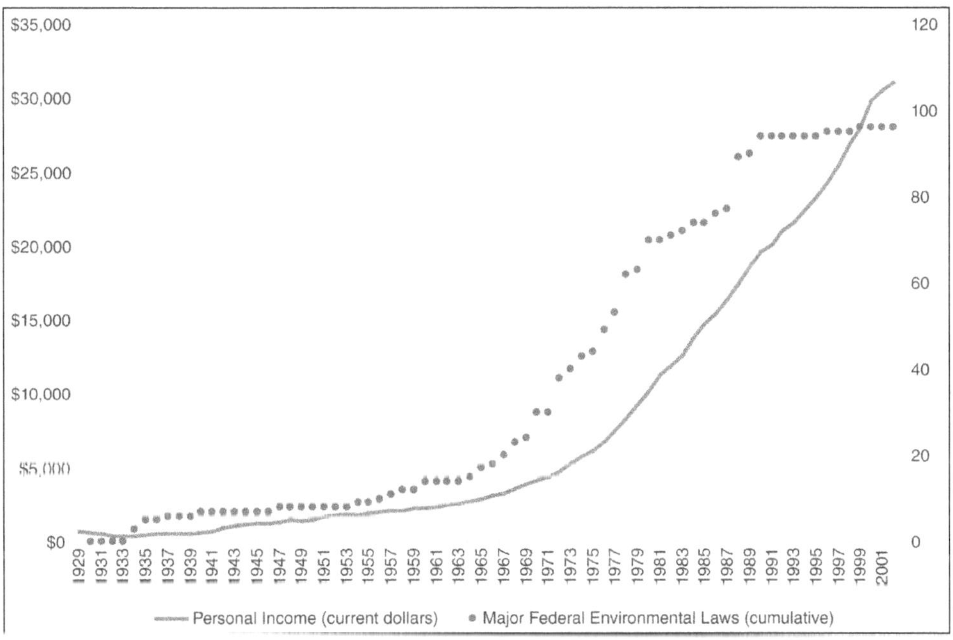

Data Sources: US Census Bureau 2009 Statistical Abstract; HDR 2001

Among those factors (and these are all arguable):

- Increased public awareness via immediate (and ever more visual) news media
- Advent of the Baby Boomer generation
- Improved laboratory analytical techniques with lower and lower detection limits (see "The Vanishing Zero" in chapter 15)
- A growing middle class with higher standards of living
- Growing social movements and accompanying activism (not necessarily related to environmental issues)
- Greater understanding of environmental epidemiology in the medical profession

The diagram on p. 27 graphs personal income (in current dollars) and the cumulative number of *major* federal environmental laws over the past 80 years. Does this prove anything? Probably not to any level of statistical rigor; but it does provide some sense that the growth in environmental laws roughly paralleled growth in individual income. Poor societies are too busy just trying to get by; it's only when a society achieves some level of well-being that it has the time, energy, and fiscal resources to invest in the environment.

The National Environmental Policy Act

One of the primary harbingers of the USEPA (and its regulations) was a groundbreaking environmental law that did not actually mandate any environmental improvements. What the National Environmental Policy Act (NEPA) of 1969 did was institute a national policy for the environment that

- Encouraged a productive and enjoyable interaction between people and the environment,
- Promoted efforts to prevent or minimize adverse environmental impacts, and
- Enriched the understanding of ecosystems and their relationship to the nation.

Perhaps the most profound impact of NEPA was its requirement to prepare a document describing any potential environmental impacts of a proposed project—an environmental impact statement or EIS. In the wake of NEPA, the USEPA was established, and the acronymic soup that pervades today's regulatory maze rapidly followed.

CHAPTER 5

THE SAFE DRINKING WATER ACT

Safe drinking water is a fact of life that we take for granted in the United States. Unfortunately, that's not the case in much of the world, where according to the World Health Organization nearly two million people die every year from diarrheal diseases (including cholera). Sadly, 90 percent of these deaths are children under the age of five, mostly in developing countries. The vast majority of diarrheal disease is attributable to unsafe water supply, inadequate sanitation, and poor hygiene.

It wasn't that long ago when waterborne disease epidemics took their toll in the United States as well. The following graph dramatically illustrates the decline in incidences of typhoid and paratyphoid fever in the United States over the past century. Typhoid fever (*Salmonella typhi*) is a common waterborne disease (although contaminated food can also cause the disease). Note that drinking water disinfection with chlorine in the United States began in 1908; this graph speaks volumes.

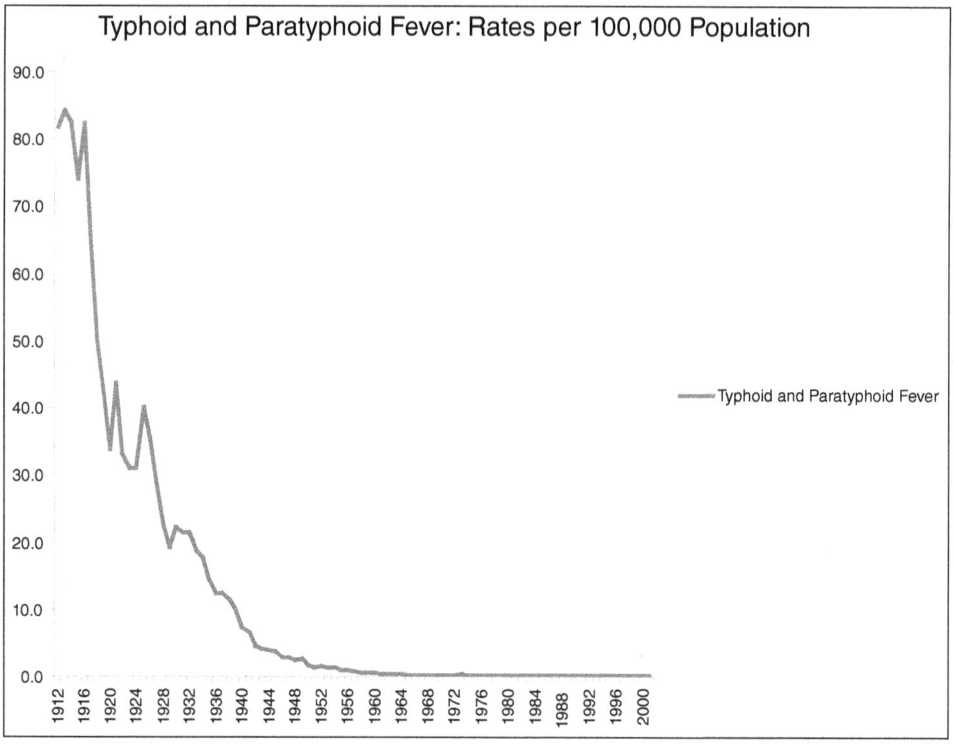

Data Source: US Centers for Disease Control

Indeed, numerous organizations, including the National Academy of Engineers and the Centers for Disease Control and Prevention (CDC), identified clean water services (drinking water treatment and disinfection, and wastewater treatment) as one of the greatest technical accomplishments of the twentieth century.

But how do we ensure that drinking water remains safe? That lofty public health goal is maintained by complying with the 1974 Safe Drinking Water Act (SDWA), which is the most important regulation affecting drinking water utilities in the United States. It impacts virtually every facet of a water utility's operations including how source waters are protected, then treated, monitored, and finally distributed to customers. Note that the only operant word in the act's title is *safe*; the act does not address adequate water quantities for fire protection or landscape irrigation—just safety for consumers.

The SDWA is the primary federal law that ensures *the quality and safety of Americans' drinking water*, which is somewhat ironic because less than 5 percent of all potable water produced is actually consumed by humans. Under the SDWA, USEPA sets standards for drinking water quality and oversees the states, localities, or tribes that implement and enforce those standards, a process known as primacy. Forty-nine states (all except Wyoming) and the District of Columbia have achieved primacy status.

The USEPA currently has established standards (known as a maximum contaminant level or MCL) for 90 chemical, microbiological, radiological, and physical contaminants in drinking water (see http://www.epa.gov/safewater/contaminants/index.html for the complete list). The agency also sets Secondary Drinking Water Regulations, which are nonenforceable guidelines for contaminants that may cause cosmetic effects (such as tooth discoloration) or aesthetic effects (such as taste or odor). Water systems are not required by the USEPA to adopt these secondary standards, but states may choose to adopt and enforce them.

The agency may also promulgate a maximum contaminant level *goal* (MCLG), which establishes the contaminant concentration at a level at which no known or anticipated adverse effects on human health occur and includes a factor of safety. The MCLG is nonenforceable and usually the starting point for determining the regulated MCL. Many MCLGs (such as for arsenic) are set at zero.

Preludes to the SDWA

Many texts report that the Egyptians used the chemical alum [aluminum sulfate or $Al_2(SO_4)_3$—still used as a coagulant] as early as 1500 BCE to settle particles out of water. Other ancient civilizations added carbon (as charcoal) to remove taste impurities. By the eighteenth century, sand filtration was known to be an effective means of removing particles from water. In 1879, a discourse on British waterworks observed, "[T]he only known methods of purifying water are distillation, oxidation (aeration), subsidence, precipitation, straining, and filtration" (Humber 1879). Mankind intrinsically seemed to know that dirty water was unsafe water, but the pathological mechanism was yet unknown.

By the middle of the nineteenth century, scientists began to gain a greater understanding of diseases and their causal factors. One particularly famous episode involved London's Broad Street Pump. In 1854, a pioneering epidemiologist, Dr. John Snow, proved cholera was a waterborne disease by linking a particularly savage outbreak of that illness in London to a public well (the Broad Street pump) that was contaminated by sewage from a nearby cesspool. Although the epidemic was abating by the time he famously removed the pump handle, his investigation was a turning point in disproving the then-consensus "miasma" (or bad air) theory of disease. Cholera was indeed a waterborne disease. (Steven Johnson's 2006 book, *The Ghost Map*, relates the story of the Broad Street pump and provides some interesting insights into nineteenth-century sanitation and recycling practices. It also details the important contributions of a local clergyman, Reverend Henry Whitehead, to this early epidemiological investigation.) In the late 1880s, progress continued with Louis Pasteur demonstrating the germ theory of disease, which explained how microscopic organisms (microbes) could transmit disease through media like water.

Dr. John Snow

When the twentieth century dawned, many large cities were treating drinking water by filtration. But as the germ theory of disease gained wider acceptance and the field of epidemiology matured, water professionals realized that drinking water would not only have to be filtered but disinfected as well, which initially occurred in the United States in 1908. But there were still no state or federal regulations governing water quality.

That changed in 1914 when the US Public Health Service issued a limited set of standards for the bacteriological quality of drinking water—but only for water systems that provided drinking water to interstate carriers such as trains (notably, one of the first specific regulations addressed the use of a shared cup for drinking water on trains). The Public Health Service revised and expanded these standards in 1925, 1946, and 1962. The 1962 standards regulated 28 contaminants and were the most comprehensive federal drinking water standards in existence before the 1974 SDWA. Interestingly, all 50 states adopted (with some modifications) these same Public Health Service standards even though there was no federal mandate to do so.

The 1974 Safe Drinking Water Act

The late 1960s and early 1970s witnessed two separate, but interrelated, trends: growing environmental awareness and activism, and improved laboratory analytical techniques that significantly lowered method detection limits (the Vanishing Zero—more on this later). As investigators sampled and analyzed source waters, they detected more and more chemicals that were suspected of causing health problems.

Contaminants Regulated Under the 1962 US Public Health Service Standards

- Alkyl benzene sulfonate
- Arsenic
- Barium
- Beta and photon emitters
- Cadmium
- Carbon chloroform
- Chloride
- Chromium
- Color
- Copper
- Cyanide
- Fluoride
- Gross alpha emitters
- Iron
- Lead
- Manganese
- Nitrate
- Phenols
- Radium-226
- Selenium
- Silver
- Strontium-90
- Sulfate
- Threshold odor number
- Total coliforms
- Total dissolved solids
- Turbidity
- Zinc

At the same time, the Public Health Service (as previously noted, in many ways a precursor to the USEPA) was looking into the condition and performance of public water works. What they found wasn't pretty. Only 60 percent of the water systems surveyed delivered drinking water that met all Public Health Service standards. Many had major structural treatment deficiencies including problems with disinfection, clarification, or distribution system pressure. Small systems serving fewer than 500 customers were especially problematic.

This nexus of growing public awareness, increasing analytical prowess, and discernible physical problems with water works led to the passage of the 1974 SDWA. The original SDWA established interim primary drinking water standards that were essentially the same as the 1962 US Public Health Service standards. The nascent USEPA also established other regulatory requirements addressing contaminant monitoring, record keeping, and public notification. Throughout 1975 and 1976, additional standards were established for synthetic organic chemicals, inorganic chemicals, total coliforms bacteria, and radionuclides with the first set of formal MCLs (not interim standards) issued in June 1977.

Interestingly, the 1974 SDWA also directed the National Academy of Sciences to conduct a study on the human health effects of various contaminants found in drinking water. However, the study was to "consider only what is required for protection of the public health, not what is technologically or economically feasible or reasonable" (HDR 2001). Keep this admonition in mind when the topics of the Vanishing Zero and emerging contaminants are discussed later in this book.

The 1986 SDWA Amendments

The 1974 SDWA was a landmark piece of legislation and established a solid regulatory framework intended to provide safe drinking water to consumers across the United States. But more regulations were on the way as Congress sought to expand both the scope of the regulation and the speed of regulatory promulgation.

Some highlights of the 1986 SDWA Amendments:

- A goal was set of 83 contaminant standards to be established by 1989 and 25 more contaminants every three years thereafter.
- Additional contaminants, including organic contaminants such as pesticides and volatile organic compounds, and disinfection byproducts (DBPs), were identified.
- A health goal (MCLG) of zero for total coliforms and an MCL for total coliforms was called for with the Total Coliform Rule.
- Lead pipes and lead solder in new drinking water systems were banned.
- States were required to develop a wellhead protection program for public wells.
- All systems using surface water (lakes or streams) as a source had to use disinfection techniques and filter their water sources to remove microbes, including bacteria, viruses, and protozoa that commonly occur in surface water under the Surface Water Treatment Rule.

Of all the new standards promulgated by the 1986 Amendments, perhaps the most far-reaching, expensive, and problematic for water utilities was the Surface Water Treatment Rule. Many of the largest metropolitan water systems used (and still use) surface water as their source of supply. Surface waters are particularly susceptible to microbial contamination from a variety of sources including point sources (such as wastewater treatment plant discharges) and nonpoint source runoff. Retrofitting older treatment plants (if possible) or building new facilities is expensive, and many large cities balked at this requirement. Several cities cited pristine and highly protected watersheds in an effort to obtain a filtration waiver. Some were successful. The Portland (Ore.) Water Bureau still does not filter its drinking water and is currently in litigation with the USEPA regarding a filtration provision in the 1996 SDWA Amendments. New York City recently relented, and a multibillion-dollar drinking water treatment plant is currently under construction.

Because of the Surface Water Treatment Rule, the Western Municipal Water District abandoned its Hardinge filtration plant located near Lake Mathews, Calif., on the Colorado River Aqueduct. This popular

1960s-vintage treatment technology did not comply with the new Surface Water Treatment Rule and was no longer an approved technology because of its shallow filter depth and lack of individual filter monitoring and filter-to-waste capability. Western turned to the then recently constructed Metropolitan Water District of Southern California Henry J. Mills water treatment plant (Riverside, Calif.) to supply finished water.

The 1996 SDWA Amendments

Ten years later, another set of major amendments brought still further changes to the maze of drinking water regulations. This time around, there was a slightly different emphasis, one that focused on *preventing* contamination problems through source water protection and enhanced water system management. These amendments transformed the SDWA from its more traditional, after-the-fact regulatory focus into a more holistic environmental statute.

One of the primary drivers of the 1996 Amendments was the 1993 cryptosporidiosis outbreak in Milwaukee, Wis. The disease is caused by ingesting *Cryptosporidium parvum,* a protozoan parasite that causes gastrointestinal illness. This episode is considered the largest documented waterborne disease outbreak in US history. It was estimated that more than 400,000 Milwaukee residents became ill and over 100 died. Many

Hardinge Filter Plant. Photograph Courtesy of Western Municipal Water District

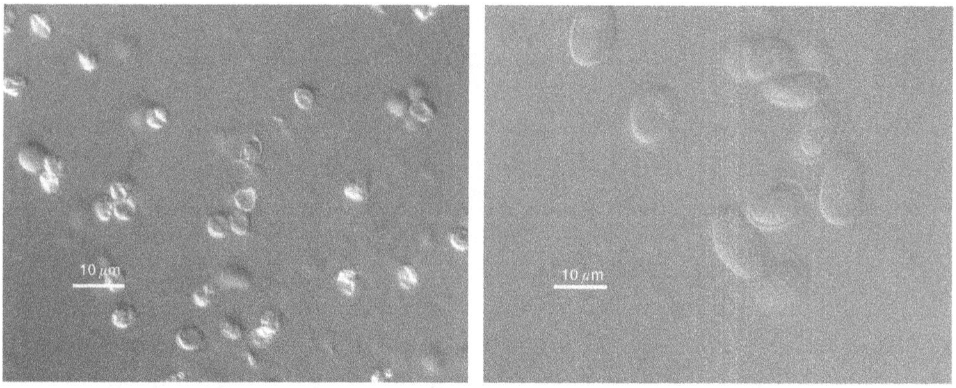

Cryptosporidium (at left) and *Giardia*. Photo Courtesy of H.D.A. Lindquist, USEPA

of the 1996 Amendments and their subsequent regulations specifically addressed *Cryptosporidium* and *Giardia lamblia* (another protozoan parasite).

Some major features of the 1996 SDWA Amendments:

- New and stronger approaches to prevent contamination of drinking water
    - ▲ Aggressive source water protection
    - ▲ Capacity development—assisting smaller water utilities with issues related to technical, managerial, and financial expertise
    - ▲ Operator certification
- Expanded "right-to-know" provisions that resulted in the annual Consumer Confidence Report
- Regulatory improvements
    - ▲ Repeal of the 10-year-old requirement mandating USEPA to regulate 25 contaminants every three years; instead, a focus on risk-based contaminant selection
    - ▲ Occurrence information on unregulated contaminants
    - ▲ Information collection
    - ▲ New contaminant regulations (promulgated via the Contaminant Candidate List or CCL process, which is currently at CCL #3)
    - ▲ Specific contaminant regulations
        - ■ Arsenic
        - ■ Radon
        - ■ Microbial/Disinfection By-Product Rules (joint promulgation)
            - ❖ Stage 1 Disinfectants and Disinfection By-Products Rule (Stage 1)

    - Interim Enhanced Surface Water Treatment Rule
    - Long Term 1 Enhanced Surface Water Treatment Rule (LT1)
    - Long Term 2 Enhanced Surface Water Treatment Rule (LT2)
    - Stage 2 Disinfectants and Disinfection By-Products Rule (Stage 2)
  - Groundwater Rule
- New funding for states and communities through the Drinking Water State Revolving Fund

There were several groundbreaking aspects to the 1996 Amendments, notably in modifying the manner in which USEPA collected information from drinking water utilities (the Information Collection Rule or ICR) and in how potential contaminants were evaluated for regulation (the CCL), as well as in its holistic support for smaller utilities (the capacity development program). But perhaps the most innovative was how the agency attacked the seemingly unrelated issues of controlling disinfection by-products (DBPs) while at the same time enhancing protection against microbial contaminants.

Recall that the 1993 Milwaukee cryptosporidiosis outbreak significantly influenced the agency's approach toward improving microbial contaminant control. More aggressive treatment—which could include disinfection—would be required to reduce microbial risk. But at the same time research was maturing on another potential public health hazard—this one caused by the very practice of water disinfection itself.

DBPs—specifically trihalomethanes, or THMs—were first discovered in the early 1970s. They are produced when disinfectants (such as chlorine) interact with naturally occurring organic materials (NOM), which are present in many surface waters. Clearly, DBPs had been around since utilities began disinfecting NOM-containing surface water with chlorine; they just hadn't been detected—or sought. DBPs were first subject to regulation in 1979 and, over the next decade, additional epidemiological research reinforced concerns about DBPs and their potentially adverse impacts to human health (cancer, and reproductive and developmental effects). It was clear that more DBP control regulations were on deck.

But there was a conundrum—health researchers imparted the need to reduce DBP exposure, but USEPA feared that reducing disinfectant dosing could adversely impact efforts to reduce microbial contamination. A regulatory negotiating (or reg-neg) committee formed under the auspices of the USEPA adopted a dual-track approach to controlling DBPs, while also focusing on ways to reduce human health risks from microbial contamination.

This was the genesis of the jointly promulgated series of Microbial/Disinfection By-Product Rules (M/DBP Rules) listed above. In a stepped series of regulations beginning with the ICR (which required large public water systems to monitor and collect data on microbial contaminants, disinfectants, and DBPs for 18 months), the Agency sought a narrow balance between controlling DBPs and reducing microbial risk. This intertwined effort included a host of new approaches to regulation including the aforementioned ICR, filter performance standards, new watershed protection requirements, and finally, a "toolbox" approach that allowed individual drinking water utilities to tailor their microbial control efforts based on risk.

USEPA and the History of the SDWA

The USEPA made a (justifiably) big deal out of the twenty-fifth anniversary of the SDWA in 1999. They touted the many advances and accomplishments achieved by the drinking water community under the umbrella of the SDWA. The agency compiled a wonderful summary of the act, its amendments, and other drinking water facts in *25 Years of the Safe Drinking Water Act: History and Trends* (USEPA 1999). It's a good introduction to the SDWA, its background, and successes; a recommended read.

A Couple of Drinking Water Footnotes

Before we move on from the SDWA to the next corridor in the green maze, there are a couple of interesting drinking water–related footnotes to cover: bottled water and commercial-aircraft drinking water.

Bottled water is *not* regulated under the SDWA; in fact, it's not even regulated by the USEPA. Because it is considered a food, bottled water is regulated by the US Food and Drug Administration (FDA), which is housed in the US Department of Health and Human Services. The relationship between drinking water purveyors and the bottled water industry is somewhat rocky at times, with both making claims about the relative purity and value (and occasionally safety) of their product vis-à-vis its competitor.

It is true that drinking water utilities are subject to more testing requirements and regulatory scrutiny than water bottlers. Similarly, whereas the SDWA requires utilities to prepare an annual CCR, the FDA makes no such demands on bottlers; they merely have labeling requirements that don't detail contaminant levels, origin of source water, or violations that might have occurred during the previous year (as the CCR requires from drinking water utilities). In fact, here's what USEPA's own FAQ web page (http://www.epa.gov/ogwdw000/faq/faq.html#bw) has to relate about bottled water:

- Bottled water is not necessarily safer than your tap water. USEPA sets standards for tap water provided by public water systems; the Food and Drug Administration sets bottled water standards based on USEPA's tap water standards.
- Bottled water and tap water are both safe to drink if they meet these standards, although people with severely compromised immune systems and children may have special needs.
- Some bottled water is treated more than tap water, while some is treated less or not treated at all. Bottled water costs much more than tap water on a per gallon basis. Bottled water is valuable in emergency situations (such as floods and earthquakes), and high quality bottled water may be a desirable option for people with weakened immune systems.
- Consumers who choose to purchase bottled water should carefully read its label to understand what they are buying, whether it is a better taste, or a certain method of treatment.

On the other hand, water stored and served on commercial aircraft *is* regulated by the SDWA as of October 19, 2009. That's when the Final Rule for aircraft public water systems appeared in the *Federal Register*. USEPA investigated the quality of drinking water on commercial aircraft from 2005 through 2008, collecting numerous water samples, which were analyzed for total coliforms (TC) and disinfectant residual. Although the eventual sample set of 12,794 TC samples was not considered statistically robust, the results weren't encouraging. Some 463 were TC positive, and of those, 18 were *E. coli* positive. Disinfectant testing wasn't much better, with nearly 3,000 samples out of a total of 16,109 registering nondetect for a disinfectant residual.

The new rule only applies to carriers operating within the United States and prescribes an Aircraft Water System Operations and Maintenance Plan, which calls for routine disinfection and quarterly flushing of onboard water storage tanks. Scheduled water sampling and analysis for TC (along with follow-up speciation for positive results) is also mandated under the rule.

CHAPTER 6

THE CLEAN WATER ACT

From one end of the pipe and the SDWA, let's move to the other end of the pipe and the Clean Water Act (CWA). Recall from the regulatory timeline that the first major US law to address water pollution was the original 1948 Federal Water Pollution Control Act. Some 25 years later, the same nexus of growing public awareness and increasing analytical capabilities that led to the SDWA also resulted in the Federal Water Pollution Control Act Amendments of 1972, more commonly referred to as the Clean Water Act.

In addition to the Cuyahoga River fire discussed previously, one cultural icon that may have contributed to the increasing public awareness of water pollution (and thus pressure on elected representatives to clean up the nation's waters) was a television commercial that debuted on the second Earth Day (April 1971) depicting an American Indian chief crying over despoiled and trash-filled waters. It was one of the most successful public service announcements ever broadcast, starring (Italian-American actor) Iron Eyes Cody as the "crying Indian."

By any measure, the CWA is one of the most important and sweeping environmental statutes ever passed. It was also one of the most controversial, and more than 35 years after its introduction it continues to be debated and litigated. Although difficult to prove one way or the other, it has been asserted that the CWA's total maximum daily load (TMDL) provision is the single most litigated environmental statute in US history. There are still ongoing lawsuits (and resulting regulatory guidance and policies) simply attempting to define the term *waters of the United States*.

Generic water pollution

The CWA's regulatory philosophy is one defined as *command and control*, which established a framework of rigorous standards and a federal enforcement mechanism (via the newly formed USEPA). The act set some ambitious goals such as achieving a zero discharge of pollutants into the nation's waters by 1985, which clearly wasn't met. Here too was related legislation to ensure that the CWA covered *all* waters of the United States (one of the provisions still being litigated today).

The first wave of CWA regulations primarily impacted municipalities and industries that directly discharged into lakes or rivers. Industries were required to install pollution control technology, and municipal publicly owned treatment works (POTWs) were to achieve no less than secondary treatment of wastewater by 1977. Provisions to control contaminated runoff or nonpoint source discharges were included in the original CWA, but promulgation of regulations and enforcement were postponed until the late 1980s and early 1990s.

There are several key provisions in the CWA:

- It established the basic structure for regulating pollutant discharges into the waters of the United States—the *National Pollutant Discharge Elimination System (NPDES)*.
- It gave USEPA the authority to implement pollution control programs such as setting wastewater standards for industries that discharged to POTWs—the *National Pretreatment Program*.
- It set *water quality standards* for all contaminants in surface waters.

- One provision made it unlawful for any person to discharge any pollutant from a point source into navigable waters, unless a permit was obtained from the Corps of Engineers (a *404 permit*).
- It funded the construction of sewage treatment plants under the *construction grants program* (known to some folks as the "golden age of wastewater consulting").
- It began planning to address *nonpoint source pollution.*

The figure below illustrates some programmatic and permit requirements under the CWA.

The National Pollutant Discharge Elimination System

The NPDES permit program is intended to control point-source direct discharges of wastewater from POTWs, industries, and other potential nonpoint sources of pollution into waters of the United States. Point sources are discrete physical conveyances such as pipes or constructed ditches. Nonpoint sources are more diffuse (such as stormwater runoff) and typically more dilute, but still can transport pollutants.

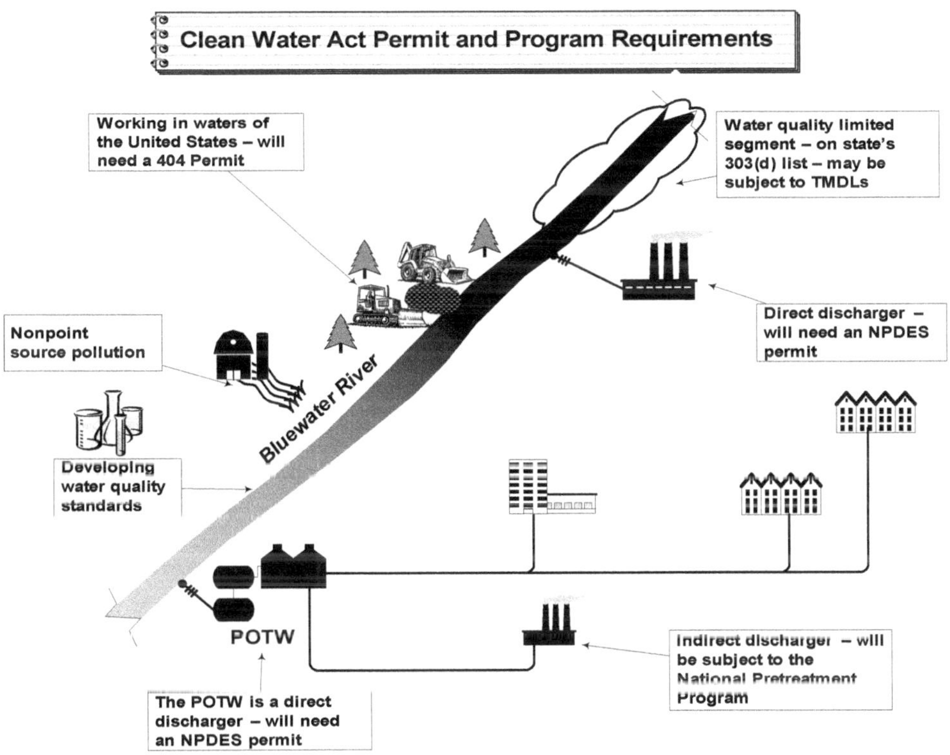

NPDES discharge permits contain numeric effluent limitations for certain pollutants (and in some cases, pollutants that are specific to certain industries). Facilities must periodically monitor their effluent (collect and analyze wastewater samples) and submit Discharge Monitoring Reports to demonstrate compliance. NPDES permits are valid for five years; upon application for renewal, the governing permit agency must provide public notice of pending permits and provide an opportunity for public comment (see the diagram on page 45).

Keep in mind that NPDES discharge permits are for direct discharges to water bodies; indirect discharges (such as from industries or commercial establishments) to a POTW via *the sanitary sewer* are addressed under the National Pretreatment Program, which is discussed in the next section.

As the NPDES program matured, its regulations expanded to include stormwater discharges. Stormwater permits typically require facilities to prepare a Stormwater Pollution Prevention Plan (a SWPPP, pronounced *swip*) and implement best management practices (or BMPs), but do not specify numeric effluent limits and may not include regular monitoring requirements (although some states require such monitoring). Today, some stormwater permits are issued as statewide general permits requiring all affected entities to abide by their monitoring and control provisions. (A section on stormwater follows.)

The flowchart on page 45 illustrates the complexity of the NPDES permit process and approximate timelines. Note that there is significant opportunity for public input and comments.

One industry that has been subject to considerable scrutiny under the CWA over the past decade-plus is animal feeding operations or AFOs. An AFO is an agricultural facility where animals (cattle, sheep, and so on.) are fed and housed in relatively confined conditions for a total of 45 days or more in any 12-month period. There are more than 450,000 AFOs in the United States but a much smaller number of concentrated AFOs (known as CAFOs), which are subject to NPDES permitting and regulations.

AFOs and CAFOs pose a potential threat to surface water quality because they can contribute pollutants such as nitrogen and phosphorus (nutrients that could lead to eutrophication of water bodies), organic matter, sediments, pathogens (such as *Cryptosporidium*), heavy metals, and, of special interest lately, hormones and antibiotics (see "Emerging Contaminants" in chapter 16).

USEPA is attempting to control runoff from these facilities by developing Comprehensive Nutrient Management Plans (CNMPs), which identify actions to meet nutrient management goals at agricultural operations. Among the components that may be found in a CNMP are:

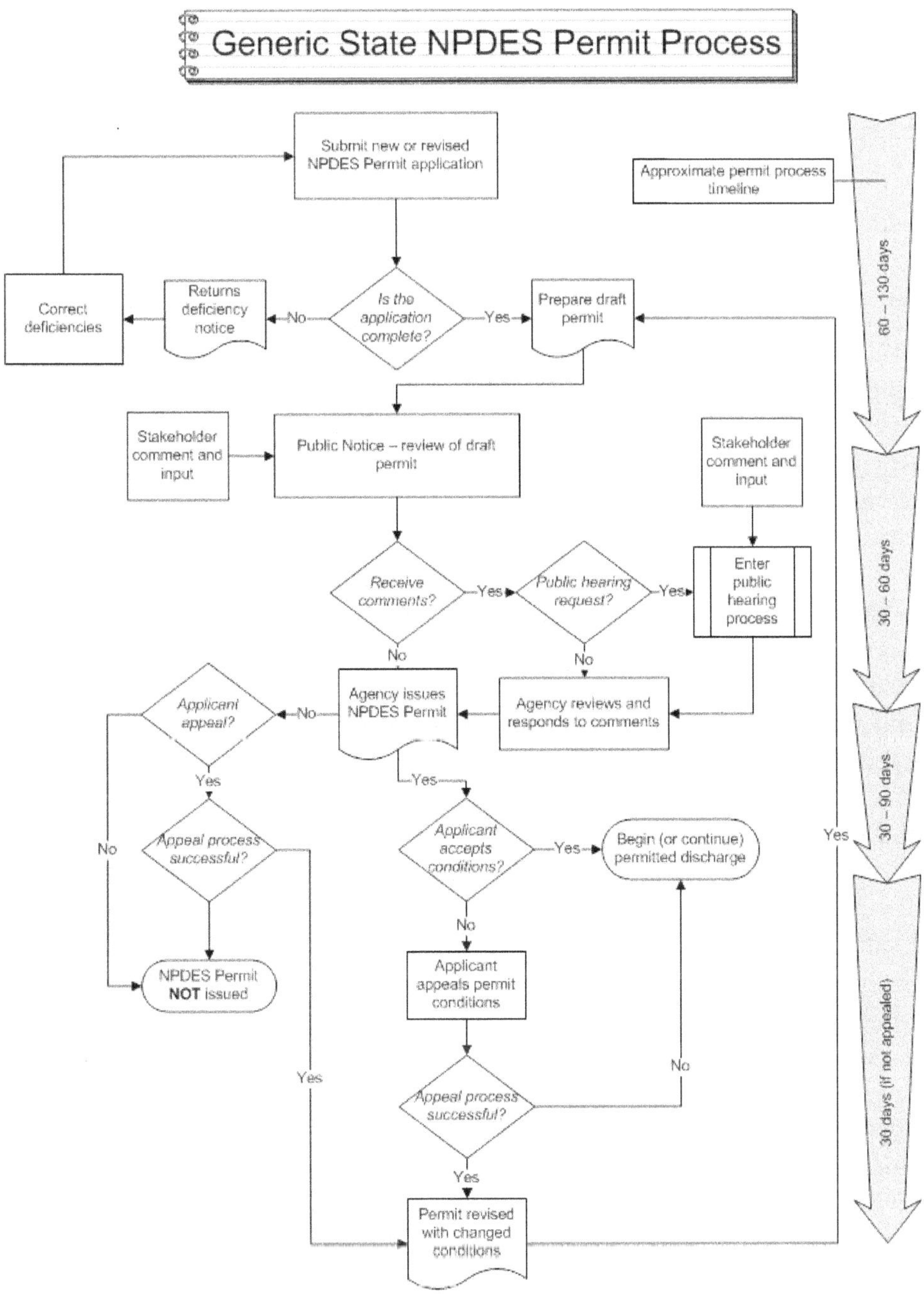
Generic State NPDES Permit Process
Submit new or revised NPDES Permit application
Approximate permit process timeline
Is the application complete?
No
Yes
Returns deficiency notice
Correct deficiencies
Prepare draft permit
Stakeholder comment and input
Public Notice – review of draft permit
Stakeholder comment and input
Receive comments?
Yes
No
Public hearing request?
Yes
No
Enter public hearing process
Agency issues NPDES Permit
Agency reviews and responds to comments
Applicant appeal?
No
Yes
Appeal process successful?
No
Yes
NPDES Permit NOT issued
Applicant accepts conditions?
Yes
No
Begin (or continue) permitted discharge
Applicant appeals permit conditions
Appeal process successful?
No
Yes
Permit revised with changed conditions
Yes
60 – 130 days
30 – 60 days
30 – 90 days
30 days (if not appealed)

- Feed management—to reduce the amount of nutrients in manure
- Manure handling and storage
- Managed land application of manure
- Other options—such as off-site management of manure, including using it for power generation

The National Pretreatment Program

In 1986, the USEPA reported that more than one-third of all toxic pollutants entering the nation's waterways did so from POTWs via industrial discharges to the sanitary sewer system. That toxic pollutants pass through a wastewater treatment plant is not surprising—these plants were never intended to treat such constituents. Conventional wastewater treatment plants are designed to treat the biological components of wastewater (biochemical oxygen demand or BOD, and total suspended solids or TSS)—not volatile or semivolatile organic compounds or heavy metals. In addition to pass-through to the receiving waters, certain uncontrolled industrial discharges can adversely impact the operations of a conventional wastewater treatment plant, damage the sewer collection system, or threaten the health and safety of utility workers.

Thus, to address these indirect discharges from industries to POTWs, the USEPA established the National Pretreatment Program as a component of the NPDES program. The Pretreatment Program permits these indirect dischargers and requires them to treat or control discharges to the sanitary sewers. It includes provisions to prevent accidental discharges from hazardous materials or waste storage areas. Although still titled the National Pretreatment Program, a more accurate descriptor of the program's intent, and one growing in use, is *source control.*

Every large POTW and many smaller POTWs (if treating wastewater from a significant industrial discharger) are required to initiate and maintain a pretreatment program to enforce all national standards. More stringent local discharge standards (known as local limits) may also be promulgated and enforced via this regulatory mechanism. Similar to the NPDES program, the pretreatment program requires industrial and commercial dischargers to obtain discharge permits and periodically monitor their effluent. Chemical constituents in the wastewater that might adversely impact the treatment plant (such as a toxic organic compound) or pass through the treatment plant and impact the receiving water (heavy metals) must be controlled via pretreatment systems. This concept of controlling contaminants that pass through conventional treatment is even more critical now that many wastewater treatment plants' effluent streams are being used as recycled water. Strict control of incoming constituents also ensures that higher-quality biosolids can be used for agricultural purposes.

Industrial pretreatment systems can reduce or eliminate a number of industrial effluent constituents including:

- Oil and grease
- Volatile organic compounds (VOCs)
- High-strength wastes (wastewater utilities often charge for excess BOD or TSS)
- Heavy metals from cleaning or plating operations
- Acidic or basic waste streams

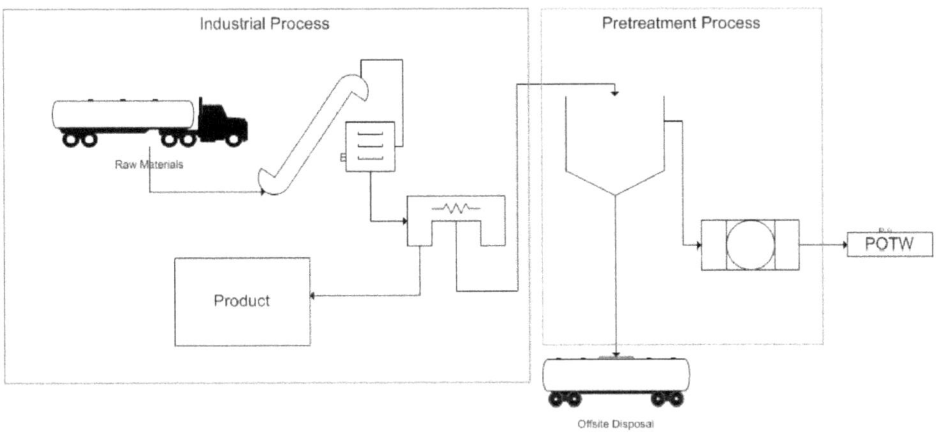

A Generic Industrial Pretreatment System

Water Quality Standards, the 303(d) Program, and TMDLs

One other important component of the CWA is the Impaired Waters and TMDL program, also known as the 303(d) program. This is a water quality program intended to achieve the desired (or designated) uses of a water such as recreation, water supply, aquatic life, or agriculture.

Under section 303(d), states (or regions such as Regional Water Quality Control Boards in California) and tribes are required to develop a list of "water quality limited segments"—waters that will not meet water quality standards for a particular pollutant (such as nitrates, metals, or total coliforms) even after technology-based permits are in place. The responsible jurisdictions must then establish priority rankings for impaired waters on these so-called 303(d) lists and develop TMDLs for these waters.

Remember that a TMDL is a total maximum daily load, which is the sum of discharge allocations to sources along the water-quality-limited segment that impair or threaten to impair that water body. Note that pollutant sources include both point sources and nonpoint sources.

TMDL = waste load allocation (for point sources) + load allocation (for nonpoint sources) = safety factor

What are water quality standards? Well, they outline the goals for that water (the designated uses noted previously), establish criteria to protect those uses, and then fashion provisions to protect water from pollutants. A water quality standard consists of four basic elements:

- Designated uses of the water body
- Water quality criteria that protect the designated uses
- An antidegradation policy to maintain and protect existing uses
- General policies addressing implementation issues

One very contentious aspect of the TMDL program is the mandate under Section 303(d) that USEPA establish, implement, and enforce TMDLs in jurisdictions that have failed to do so to the satisfaction of either the USEPA or an aggrieved plaintiff (another one of those stakeholders). One indication of the litigious nature of the TMDL program is the fact that the USEPA dedicates an entire Web page to the history and status of litigation under the 303(d) program (http://www.epa.gov/owow/tmdl/lawsuit.html).

The following diagram illustrates the basic 303(d) and TMDL process. Like so many other graphic representations of a regulatory process, it is not static. It represents an ongoing and dynamic regulatory negotiation process between regulators and the regulated community, which includes determinations of the constituents to be included on a 303(d) list, and how the TMDLs (or waste load allocations) will be implemented and enforced. Including a new constituent on a water body's 303(d) list can result in the requirement for industries and municipalities alike to invest millions of dollars in new treatment technologies and monitoring programs.

Stormwater Regulations

As mentioned in the previous section, the NPDES program was always intended to deal with point source runoff, including stormwater. Uncontrolled stormwater runoff can introduce a host of pollutants into waterways including silt, heavy metals, toxic organic compounds, and oil and

Water Quality Standards, the 303(d) Program, and TMDLs

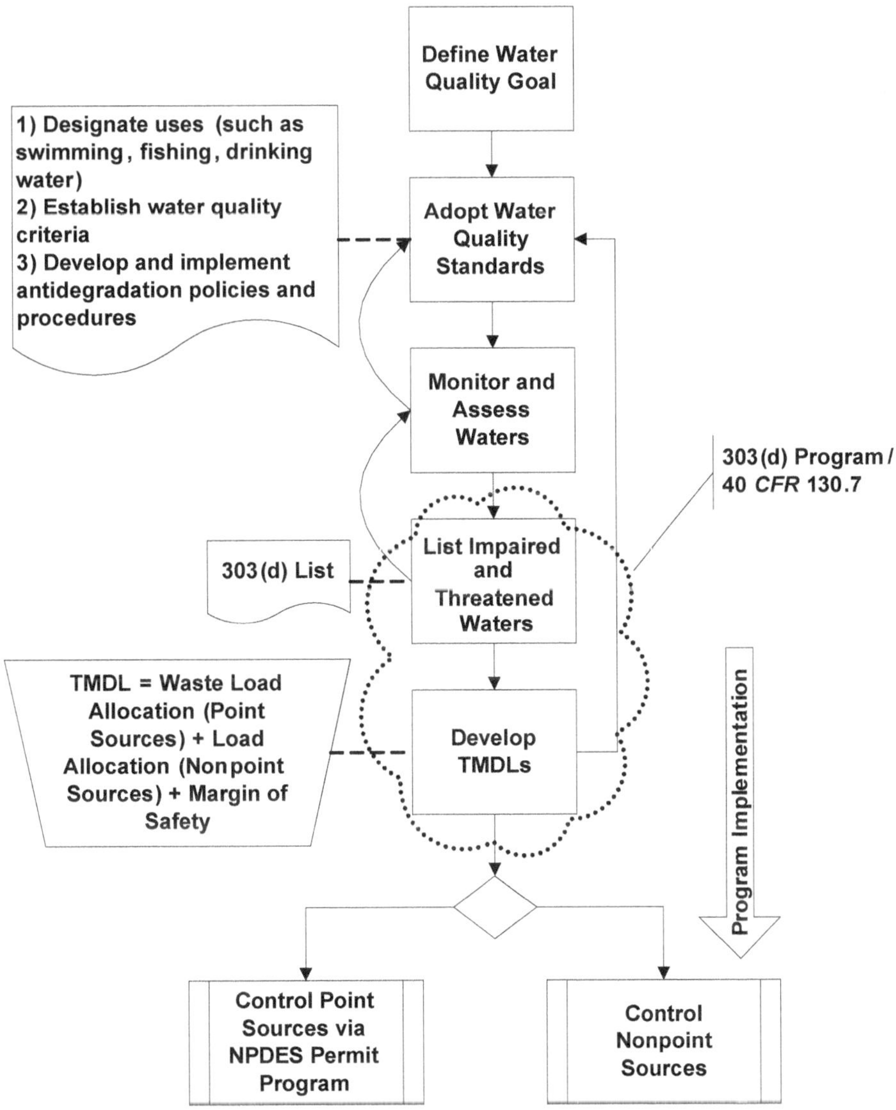

Adapted from USEPA

grease. As you can imagine, addressing and controlling the diluted—but still potentially impacting—contaminants in stormwater runoff is more difficult than controlling those in a concentrated waste stream discharged into a discrete conveyance such as a pipeline.

Thus, USEPA's initial regulatory focus following passage of the CWA was on controlling discrete point sources of water pollution (again from industrial facilities and POTWs). It wasn't until 1990 that the NPDES stormwater program was established. Phase 1 of those regulations required an NPDES permit for stormwater runoff from:

- Large or medium municipalities with populations ≥ 100,000
- Construction sites disturbing more than 5 acres
- Companies (primarily industrial) in 11 specific categories

Phase 2 was enacted nine years later in 1999. It expanded on the Phase 1 regulations by bringing smaller communities into the regulatory mix, and introducing a stormwater management program.

The MS4 Program

In urban areas, stormwater runoff is typically captured and transported via Municipal Separate Storm Sewer Systems (MS4s, but in layman's terms, simply storm sewers). In the past, this untreated runoff was often discharged directly into local water bodies. Now, to prevent the dilute (but again remember potentially impacting) pollutants from entering the water bodies, MS4 operators must obtain an NPDES permit and develop a stormwater management program.

The Phase 1 permit program actually mandates a formal NPDES permit, while the Phase 2 program enrollees noted previously are typically covered by a general permit, which requires that BMPs be employed to mitigate or control stormwater runoff and minimize or prevent the introduction of pollutants into watercourses. I'm sure you've seen examples of numerous BMPs at construction sites: they include plastic silt fences, sandbags, and hay bales.

Stormwater regulations continue to evolve, with USEPA recently issuing new effluent limitation guidelines and new source performance standards (NSPS) for construction sites. This November 23, 2009, regulatory action established technology-based minimum requirements for erosion controls that will apply nationally. The rule will eventually usher in numeric limitations on pollutants discharged from construction sites, with larger (> 20 acres) construction required to monitor discharges—the first time USEPA has imposed such national monitoring requirements and enforceable numeric limitations on construction-site stormwater discharges.

Structural Stormwater BMP—a Silt Fence.
Photograph courtesy of USEPA

Traditionally, flood control and stormwater engineers attempted to channelize and transport storm flows to the nearest waterway as quickly as possible, usually via large concrete storm channels or storm sewers. That philosophy is changing as water resource engineers realize that stormwater is yet another potential source of drinking water. By using BMPs to intercept pollutants and debris and then directing stormwater to infiltration or spreading basins, stormwater can be used to replenish or recharge falling or depleted groundwater aquifers. Care has to be taken that such actions do not interfere with senior water rights in those states governed by prior appropriation doctrine.

This more holistic "total water" approach to stormwater control (and capture) has led to a number of innovative devices and site designs for improving stormwater quality, including new filter media, packaged stormwater filtration appurtenances, hybrid filtration systems using vegetation, riparian buffers, and wetland buffers. Depending on their characteristics, these devices and systems can remove silts and suspended solids, nitrogen, phosphorus, bacteria, and even some heavy metals.

Nonpoint Sources

Although stormwater runoff is most assuredly a nonpoint source (in that it is sheet runoff), in MS4s, CAFOs, or certain industrial/commercial sites it is considered a *regulated* point source requiring an NPDES permit. There are many true nonpoint sources that can be equally impacting to surface water quality, which USEPA does not regulate. Nonetheless, the agency remains concerned about these pollutant sources as well. These unregulated nonpoint sources are also caused by rainfall or snowmelt runoff moving over and through the ground. And because the stormwater doesn't know whether it's regulated or not, it picks up and transports natural and anthropogenic pollutants into surface water bodies; pollutants can include:

- Excess fertilizers, herbicides, and insecticides from agricultural lands and residential areas
- Oil, grease, and toxic chemicals from urban runoff
- Sediment from improperly managed sites including construction sites, farmland, and eroding natural features
- Salt from roadways and irrigation practices
- Bacteria and nutrients from livestock, pet wastes, and faulty septic systems

USEPA attempts to control nonpoint pollution sources through a variety of community outreach programs, volunteer groups, and local efforts. A 1987 amendment to CWA Section 319 (Nonpoint Source Management Program) allows states, territories, and tribes to receive grant monies to support outreach programs that include technical assistance, education, training, and demonstration projects.

An example of one such local group is the Arkansas River Coalition (ARC), which is a nonprofit organization working to protect and improve the Arkansas River Watershed (http://www.arkriver.org/). ARC's Web site notes that the organization promotes "environmental stewardship through activities such as river cleanups and water quality monitoring events." Its stated goals are to:

- Protect water and the natural ecology of the river and its tributaries
- Promote recreational opportunities and the development of natural areas for public use
- Participate with individuals; local, state, and federal government; and other interested agencies and organizations to meet our goals
- Raise awareness of the history and value of the river
- Collaborate with other conservation groups within the watershed

Regulatory Intersection of the CWA and the SDWA

The USEPA recently embarked on a new rulemaking process to address direct and indirect discharges of drinking water treatment residuals to surface waters. Entitled "Drinking Water Treatment Effluent Guidelines," the rule may affect drinking water treatment plants of all sizes that discharge suspended solids, aluminum/iron salts (coagulants), organic matter, radionuclides (from source waters), polymers, lime (from softening processes), arsenic (from source waters), desalination concentrates (such as reverse osmosis brine streams), or other residuals. At the time of this writing (2010), the agency is in the midst of gathering data from drinking water utilities across the United States about the nature and quantity of treatment residuals that are directly discharged to surface waters. Survey letters have been sent to utilities requesting information on:

- Nature of treatment residuals (for example, liquid, slurry, sludge, or solid)
- Analytical data
  - ▲ Source water
  - ▲ Treatment chemicals
  - ▲ Thickening agents
  - ▲ Residuals
- Discharge Monitoring Reports (remember, they're part of the NPDES program reporting requirements)
- Current residual management/disposal practices and options
- Other operational issues that may affect the rulemaking

Stay tuned: the green maze just gets more complex.

CHAPTER 7

THE CLEAN AIR ACT

When most people hear the words *air pollution* they usually think of the term *smog* (smoke + fog), which was allegedly coined over one hundred years ago by a British antismoke activist (think current environmental activist) named Dr. H. A. Des Voeux. The good doctor was fuming (no pun intended) against the infamous London "pea-soup" fogs, which are considered classical or reducing smogs. These are produced by sulfur dioxide and particulate emissions from soft-coal combustion. Although featured in many old black-and-white Sherlock Holmes movies, pea-soupers are no laughing matter. A 1952 pea-soup smog episode resulted in between 1,800 and 2,800 deaths in the greater London area. Modern power plant emission control systems have largely reduced the frequency and impact of classical smog. On the other hand, the smog found in southern California and other major urban centers is (typically) a photochemical smog produced by combining volatile organic compounds, nitrogen oxides, and sunlight (energy) to produce ground-level ozone.

Images of billowing industrial smokestacks weren't always considered a bad thing. At the end of the nineteenth and into the mid-twentieth century, such images were iconic, indicating the vitality of the industry, and indeed, of the nation itself. Industry meant technology, and technology meant progress, and wasn't technology going to solve *all* our problems? The image below illustrates just how proud Columbia Bicycles (a particularly "green" form of transportation) was of its factory. Can you image such an advertisement today from any corporation?

Congress passed the modern Clean Air Act (CAA) in 1970. The 1970 act built on the 1963 act, which was the first major federal legislation regarding air pollution control (earlier 1955 and 1959 air pollution acts merely provided federal funds for research and local/state technical assistance). Much like the SDWA, this early federal program was administered by the US Public Health Service. The 1970 CAA significantly expanded the federal government's role in air pollution control, authorizing development of comprehensive federal and state regulations to limit emissions from both stationary (industrial) sources and mobile sources. The legislation also expanded federal enforcement authority, which was now housed

The Columbia Bicycle Factory.

Billowing smokestacks: A sign of industrial health 100 years ago.
Image courtesy of Ken Kowal at http//:www.columbiavintagebikes.com

within the newly established USEPA. Four major regulatory programs emerged from the 1970 CAA:

- National Ambient Air Quality Standards (NAAQS)
- State Implementation Plans (SIPs)
- New Source Performance Standards (NSPS)
- National Emission Standards for Hazardous Air Pollutants (NESHAPs)

Major amendments to the CAA were passed in 1977 and 1990. The 1977 Amendments primarily addressed air quality in areas of the country not meeting one or more federal air quality standards (known as nonattainment areas—for example, the southern California basin is an extreme nonattainment area for ground-level ozone, and a serious nonattainment area for both particulate matter and carbon monoxide). The 1990 Amendments significantly expanded the authority of the USEPA to address local air pollution and introduced regulations addressing acid rain and stratospheric ozone protection (which regulated, and in some cases eliminated, chlorofluorocarbon compounds). Both the NESHAP and NAAQS programs were modified and expanded.

Under the CAA, the USEPA administers a number of regulatory programs. SIPs outline how states (or in some cases, regions) intend to comply with federal air quality requirements. Those requirements are known as NAAQS and regulate pollutants considered harmful to public health and the environment. There are two types of national air quality standards: 1) primary standards, which establish limits to protect public health, including the health of sensitive populations such as asthmatics, children, and the elderly; and 2) secondary standards that protect public welfare (protection against decreased visibility; damage to animals, crops, vegetation, and buildings). NAAQS have been established for:

- Carbon monoxide (CO)
- Lead (toxic)
- Nitrogen dioxide (NO_2, a precursor of photochemical smog; you may also see the abbreviation NOx, which is all oxides of nitrogen)
- Particulate matter (coarser, known as PM_{10})
- Particulate matter (finer, known as $PM_{2.5}$)
- Ozone (O_3)
- Sulfur dioxide (sulfur dioxide [SO_2] reacts with water vapor to create sulfate ions, the precursors to sulfuric acid and acid rain)

As the next graph dramatically illustrates, the United States has made considerable progress in reducing air pollutants over the past 35-plus years (below). Nonetheless, USEPA estimates that over 120 million people live in counties that still exceed one or more NAAQS.

Of course, numerous other hazardous air pollutants or HAPs such as benzene, ethylene oxide, mercury compounds, and radionuclides are regulated under the NESHAP program, which is codified in CAA Section 112. The CAA requires a preconstruction new source review (NSR) of major new facilities to determine their potential impacts on air quality. They are then subject to NSPS and may be required to install the best emission control equipment possible, known as maximum achievable control technology or MACT. Note that these technology-based standards are only issued for major sources (and certain other types of area sources). Major sources are those sources that emit or have the potential to emit 10 tons per year or more of a HAP or 25 tons per year or more of a combination of HAPs.

USEPA administers three major permit programs under the CAA:

- Acid rain permits (Title IV permits)
- Preconstruction NSR permits
- Operating permits (Title V permits)

The Title IV program aims to reduce SO_2 emissions (a precursor of acid rain) from power plants through a permitting program coupled with an emissions trading program. The NSR program requires that stationary sources obtain permits to construct new (or modified—and this is a highly contentious point under NSR) emission sources or emission control systems. NSR permits are typically issued by local or state regulatory

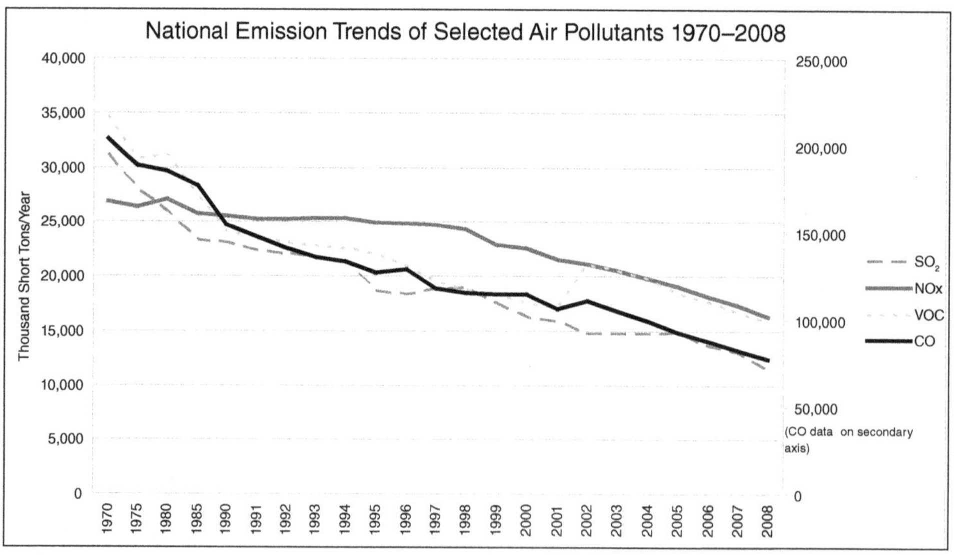

Data Source: National Emissions Inventory

agencies. After construction under the NSR permit, the major stationary source (and a limited number of smaller or area sources) must obtain an operating or Title V permit, which lists legally enforceable operating conditions including prescribed monitoring, emission rates, and required maintenance. These operating or Title V permits are also usually issued by state or local regulators.

Risk Management Plans

One other key component of the CAA is Section 112(r) and its implementing regulation, the Risk Management Plan Rule or RMP Rule. Similar to the Emergency Planning and Community Right-to-Know Act (see the section Hazardous Materials Management and Right-to-Know Regulations on page 64), the RMP Rule is aimed at reducing the risk from extremely flammable or toxic chemicals at the local level.

Facilities—including water and wastewater utilities—that use specified flammable and toxic substances are required to develop a risk management program, which includes such features as 1) a hazard assessment to evaluate the potential impacts of an accidental chemical release, 2) an aggressive safety and hazard prevention program, and 3) preparation of an emergency response program. The RMP Rule further requires that covered entities submit a summary of their entire program wrapped up in a document known as a Risk Management Plan, the contents of which are publicly available.

The chemical most likely to trigger the applicability of this regulation to water and wastewater utilities is chlorine. The threshold quantity (specified by the RMP Rule) is only 2,500 pounds. Larger water and wastewater treatment plants that store and use more than this amount of gaseous chlorine for disinfection purposes are subject to the rule.

Utilities and the Clean Air Act

Aside from the RMP Rule, why should drinking water and wastewater utilities be concerned with the CAA? Although utilities may be more affected by local air pollution control regulations (such as those issued by southern California's South Coast Air Quality Management District or SCAQMD), emissions from wastewater treatment plants are specifically regulated under federal NESHAP regulations. There are also NSPS for wastewater sludge incinerators and new digester gas-fired boilers larger than 10 million Btu per hour.

It's fair to say that generally, only the largest facilities are impacted by the federal CAA, but as mentioned, in some areas of the country (typically large urban areas, most notably southern California) permits are required for relatively small pieces of equipment including stationary natural gas engines (at pump stations or lift stations), portable diesel-fired generators, and boilers. In California, the federal CAA regulations and the attempt to comply with NAAQS are administered (in a much expanded and more rigorous fashion) by the California Air Resources Board (CARB) and various Air Quality Management Districts (AQMD).

In many regions of the country, the federal CAA is typically regulated by state-level agencies such as the Texas Commission on Environmental Quality (also responsible for water pollution and solid waste). Other states, such as Iowa, Kansas, and Missouri, share CAA enforcement responsibility between a state agency and local or county-level regulatory agencies. Determining which regulatory authority is responsible for CAA enforcement (and the other environmental statutes as well) is one of the first checklist items for anyone entering the green maze.

I believe that understanding and complying with federal or local air quality regulations is the most difficult regulatory challenge facing any industry—including water and wastewater utilities—today.

CHAPTER 8

HAZARDOUS WASTE AND HAZARDOUS MATERIALS REGULATIONS

What is hazardous waste? Well, a significant portion of federal hazardous waste regulations (notably 40 *CFR* 261) is concerned with answering just that question—identifying hazardous waste. Believe me, it can get tricky. In fact, the history of hazardous waste regulations and complying with that extremely complex corner of the green maze could be—and is—the subject of entire courses and textbooks. One of the best is *The Complete Guide to Hazardous Waste Regulations: RCRA, TSCA, HSWA, EPCRA, and Superfund*, 3rd Edition, by Travis P. Wagner.

Back to the first question: what is hazardous waste? OK, hold on and work with me: first it has to be a solid waste (and there's a specific definition for that), which has been discarded (meaning abandoned [disposed of, burned, incinerated, or stored]), and is either listed by USEPA (or a state) as hazardous or exhibits a defined *hazardous waste characteristic* such as being corrosive, ignitable, reactive, or toxic. Not so complicated, eh?

Now that the definition is out of the way, just why should water and wastewater utilities be concerned with hazardous waste regulations? Depending on where the utility is located, and the nature of its operations, there's a better than even chance that it is a hazardous waste generator (that needs a permit) that has to transport its waste off-site (via a licensed hazardous waste transporter), to a permitted hazardous waste treatment or disposal site.

In plain language, a hazardous waste is a substance that if managed or disposed of improperly could cause acute or chronic harm to human health or the environment. That's a pretty straightforward and accurate definition. Examples are heavy metals sludge from plating operations, spent acidic solutions (with a pH less than 2.0), or spent halogenated hydrocarbons. Improperly managed/disposed of, they could cause acute (immediate) impacts to human health (such as acid burns) or chronic (long-term) impacts by introducing carcinogenic compounds into groundwater.

USEPA currently divides hazardous waste into three broad categories:

- Listed wastes (defined and listed as hazardous by USEPA)
  - ▲ F-list wastes from nonspecific source wastes generated by common industrial manufacturing processes
  - ▲ K-list wastes from source-specific processes
  - ▲ P-list and U-list wastes from discarded commercial chemical products
- Characteristic wastes
  - ▲ Ignitable wastes
  - ▲ Corrosive wastes are acids or bases (pH less than or equal to 2, or greater than or equal to 12.5)
  - ▲ Reactive wastes, unstable under normal conditions and capable of causing explosions, toxic fumes, gases, or vapors when heated, compressed, or mixed with water
  - ▲ Toxic wastes, which are harmful or fatal when ingested or absorbed
- Universal wastes
  - ▲ Batteries
  - ▲ Pesticides
  - ▲ Mercury-containing equipment
  - ▲ Bulbs (lamps)

History of Hazardous Waste Regulations and RCRA

Much like the SDWA, CWA, and CAA, there were predecessors to the USEPA-enacted hazardous waste regulations of the 1970s. The 1965 Solid Waste Disposal Act was the immediate precursor to modern hazardous waste legislation, and it was the first piece of federal legislation to address the hazardous waste management problem. Significant amendments to that act in 1976 resulted in the Resource Conservation and Recovery Act (RCRA), which is the basis for today's hazardous waste management scheme. Although enacted in 1976, RCRA regulations weren't finalized until 1980. RCRA was closely followed by the Hazardous and Solid Waste Amendments of 1984 (HSWA). Catchy titles, eh? Not at all straightforward like Clean Air Act or Clean Water Act.

The legislative intent of RCRA was to properly manage hazardous wastes from cradle to grave. The USEPA is charged with establishing minimum acceptable standards for all aspects of hazardous waste generation, storage, transportation, and treatment/disposal, and permits are required for hazardous waste generators; transporters; and treatment, storage, and disposal (TSD) facilities. But despite these permit requirements, the generator retains ultimate liability for its hazardous waste and must ensure that USEPA's minimum acceptable standards are met at each stage; that's the concept behind cradle to grave (responsibility).

USEPA's hazardous waste regulations also address such topics as waste minimization, hazardous waste recycling, corrective action (under RCRA, not Superfund, which is described next), hazardous waste test methods (compiled in SW-846, *Test Methods for Evaluating Solid Waste,*

This is *not* proper hazardous waste management.

Physical/Chemical Methods), and international waste issues. As you can imagine, SW-846 is an enormous document, but is now offered in its entirety online at http://www.epa.gov/epawaste/hazard/testmethods/sw846/index.htm.

Superfund

It's a bird! It's a plane! No, it's Superfund. Probably not terribly humorous (or at this point, original) considering that the Comprehensive Environmental Response, Compensation, and Liability Act (CERCLA) or Superfund law was intended to address abandoned or uncontrolled hazardous waste disposal sites. Rather hastily written and passed in the waning days of the Carter Administration, the law was a result of a federally declared state of emergency that allowed the USEPA to respond and secure the infamous Love Canal site. The high visibility of this incident and resultant public pressure to "Do something!" led to CERCLA, which authorizes federal action to remediate, or to require others to remediate, uncontrolled hazardous waste sites. Responsibility for the cleanup is apportioned out to potentially responsible parties or PRPs. Given the enormous costs involved in remediating uncontrolled hazardous waste sites, it's not surprising that CERCLA liability is often the subject of lengthy and expensive litigation.

There are a host of other regulatory requirements under both RCRA and CERCLA, most notably that of spill notification (under yet another acronym, known as EPCRA), which is covered in the following section.

Hazardous Materials Management and Right-to-Know Regulations

Around midnight on December 3, 1984, the Union Carbide plant in Bhopal, India, accidentally released about 42 tons of methyl isocyanate, which is an extremely toxic hazardous material used as a feedstock for producing carbamate pesticides. It's estimated that more than 500,000 people were exposed to this gas with an immediate death toll of more than 2,000. Other authorities cite near-term deaths of 8,000–10,000 within 72 hours. Public health officials estimate some 25,000 have since died from gas-related conditions. The proximate cause of the leak is known (water entered a tank and the resulting chemical reaction produced over-pressures, eventually resulting in the gas leak), but the underlying or root cause remains controversial, with some (particularly Union Carbide) maintaining employee sabotage, and others pointing to unsafe conditions, poor maintenance, and improperly operated safety systems.

One of the consequences of this horrible tragedy was a greater regulatory focus not on *hazardous waste*, but on the use and storage of *hazardous materials.* A scant two years after the Bhopal disaster, Congress passed the Emergency Planning and Community Right-to-Know Act of 1986 (EPCRA), commonly known as SARA Title III (because it was part of the Superfund Amendments and Reauthorization Act). It established requirements for federal, state, and local governments, and industry, regarding emergency planning and "Community Right-to-Know" reporting for hazardous materials. The regulations pointed in two different but related directions: one at local officials and first responders dealing with the proper identification, management, and storage of hazardous materials; the second at the community intending to provide information about the chemicals in use at (and released from) facilities in their neighborhood. In short, the regulation was about providing timely and accurate information about hazardous materials in use or stored at a facility.

There were several key provisions in EPCRA:

- *Sections 301 to 303. Emergency Planning* required local governments to prepare chemical emergency response plans, and to review plans at least annually.
- *Section 304. Emergency Notification* required facilities to immediately report accidental releases of hazardous substances in quantities exceeding certain defined threshold quantities known as reportable quantities (RQs).
- *Sections 311 and 312. Community Right-to-Know Requirements* mandated that facilities manufacturing, processing, or storing certain hazardous materials maintain and provide Material Safety Data Sheets (MSDSs) describing the properties and health effects of these chemicals to state and local officials and fire departments. Complying with Section 312 results in a facility producing a Hazardous Materials Business Plan, which details the amount and nature of hazardous materials stored on site, their locations, and release prevention measures (such as secondary containment).
- *Section 313. Toxics Release Inventory (TRI)* required facilities to submit an annual Toxic Chemical Release Inventory Reporting Form for each of the more than 600 chemicals that are manufactured or otherwise used above the applicable threshold quantities.

The federal EPCRA regulations are typically enforced at the state or local level by State Emergency Response Commissions (SERCs) or Local Emergency Planning Committees (LEPCs). LEPCs consist of emergency management agencies (such as the California Office of Emergency Services), first responders (fire and police), industry, and the general public.

Asbestos-containing materials . . . sort of, kind of hazardous waste . . . but not really . . .

The production, use, and disposal of asbestos and asbestos-containing materials (ACM) is not regulated under RCRA, but rather under the Toxic Substances Control Act (TSCA), which also regulates polychlorinated biphenyls (PCBs) and lead-based paint. Because the primary exposure route of asbestos is airborne, required controls during demolition, removal, or disposal are covered by CAA regulations found at 40 *CFR* Part 61, Subpart M. Water and wastewater utilities are most likely to encounter asbestos in older asbestos–cement pipe or transite. Transite pipe manufactured today no longer contains asbestos.

Water/Wastewater Utilities and Hazardous Waste

Let's get back to the earlier question—why should water and wastewater utilities be concerned with hazardous waste regulations? Old water pipes used lead compounds in joints, and many older service lines were manufactured of lead; if you're disposing them, they're hazardous waste. Spent acidic cleaning solutions (say for membranes) with a pH less than 2.0 can't just be poured down the drain: that would violate both the CWA and RCRA; it's a hazardous waste. How about sodium hydroxide used to adjust finished-water pH for corrosion control—what if a tank becomes contaminated or out of spec? Again, you can't just toss it in the dumpster; it's a hazardous waste. Old transite pipe contains asbestos, which although technically not a RCRA hazardous waste, requires special handling and disposal procedures. So if your utility replaces and disposes of transite pipe, check out the sidebar, above.

Luckily for public utilities, CERCLA specifically exempts state and local governments from CERCLA liability. Thus, unless gross or willful negligence is involved, public utilities are not responsible for costs or damages resulting from a response to an emergency release of a hazardous substance. Nonetheless, when acquiring land for facilities, utilities should be aware of the potential for adverse impacts to the site from past misuse, storage, or disposal of hazardous materials or waste.

CHAPTER 9

RECYCLED WATER

All water on earth is essentially recycled: water is always in motion, and the water cycle (the hydrologic cycle) describes this continuous motion of water on, above, and below the surface of the earth, and it is a cycle—with no beginning or end. Sometimes these processes occur very rapidly; others take place over millions of years. There is no "new" water: the glass of water you drank yesterday may have fallen as rain in India last year or have been part of an unnamed inland sea 100 million years ago. The US Geological Survey maintains a great Web site on natural water recycling at http://ga.water.usgs.gov/edu/watercycle.html.

But that's not the kind of water recycling this chapter is going to cover; we're going to discuss *recycled water*: water which, as a result of treatment of waste, is suitable for direct *beneficial use*. Depending on the degree of treatment, that beneficial use can be:

- Irrigation applications
  - ▲ Turf (parks, playgrounds, schoolyards, golf courses)

    - ▲ Nursery stock
    - ▲ Certain types of food crops and trees
- Industrial cooling water or boiler feed
- Soil compaction/dust control
- Fire fighting or fire protection systems
- Groundwater recharge

Depending on the state or jurisdiction, water recycling can also be termed *water reuse* or *water reclamation* (with the associated term *reclaimed water*). For the balance of this chapter, the term *recycled water* will refer to highly treated wastewater that is beneficially used.

What Recycled Water Is

Because California is a pioneer in recycled water use, many other states have adopted its terminology (and regulatory scheme). In fact, as early as 1960, California legislation encouraged "wastewater reclamation and reuse," which was codified in the state water code. Current California regulations specify two basic grades of recycled water that can be produced

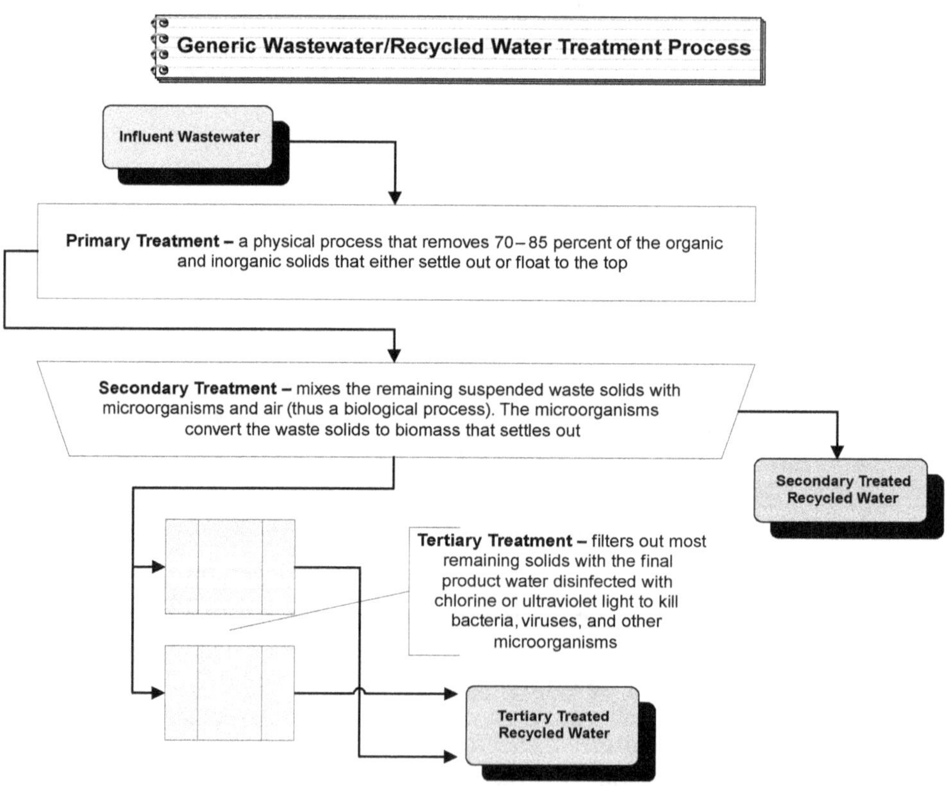

from the wastewater treatment process: secondary and tertiary. Tertiary-treated water undergoes a final filtration step and is considered a higher-quality recycled water that can be used with fewer restrictions. Recycled water use in California is regulated by the California Department of Public Health (22 *CCR* Division 4 Chapter 3); collectively, these regulations are known informally as "the purple book" (because of the requirement to identify all recycled water piping and appurtenances with a purple color). It is a mature regulatory process with established guidelines and procedures. Recycled water projects must receive formal approval through a permit process. The regulations consider unknown risks and require the use of multiple barriers of treatment, monitoring, and compliance with specific standards. The diagram on the opposite page illustrates a generic wastewater treatment/recycled water process.

Recycled Water Uses

OK, so Mother Nature has been recycling water for billions of years; people have only been formally recycling water since the nineteenth century, which is when the first incidences of modern wastewater recycling were recorded (however, untreated wastewater has been used for agricultural irrigation throughout history). By the first decade of the twentieth century, an estimated 35 communities in California were using untreated or minimally treated wastewater for farm irrigation. In 1912, recycled water (of an unknown quality) was used for landscape irrigation in San Francisco's Golden Gate Park. As the twentieth century progressed, more advanced wastewater treatment technologies (primarily the activated sludge process) and chlorination led to more widespread use of recycled water.

The modern era of recycled water use began in the 1960s when the City of Irvine, Calif., installed a dual distribution system to deliver water for both irrigation and industrial uses. In 1962, the County Sanitation District of Los Angeles County (Calif.) began planned groundwater recharge with secondary treated wastewater using spreading basins at the Montebello Forebay. In 1976 witnessed the Orange County Water District's Water Factory 21 and its direct injection of recycled water to act as a seawater intrusion barrier and groundwater recharge in Fountain Valley, Calif.

By far, the most widespread use of recycled water in the United States is for irrigation. The massive reference work *Water Reuse* (Metcalf & Eddy 2007) notes that more than 60 percent of recycled water in California and Florida is used for agricultural and landscape irrigation. Similar irrigation use percentages can probably be expected in other states. Notably, at 9 percent, groundwater recharge is the third most common use of recycled water in those two states.

Federal Recycled Water Regulations

Believe it or not, there are *no* federal regulations governing recycled water production, quality, or use! None. There are, however, recommended federal guidelines addressing a variety of reuse applications that were published by the USEPA in *Guidelines for Water Reuse* (USEPA 2004). It reports that 20 states have adopted formal regulations, 16 issue guidelines or design standards, and 9 have no regulations or guidelines but reuse may be permitted on a case-by-case basis. About 90 percent of water reuse occurs in four states: Arizona, California, Florida, and Texas. Thus, it is not surprising that the most detailed and rigorous regulations are found in these states. California's are perhaps the most well known and have been extensively used as a basis for other states' regulations.

Guidelines for Water Reuse is a valuable resource and recommended reading for any utility that is planning a recycled water program. It addresses a number of critical issues, including 1) appropriate recycled water applications, 2) technical issues (sources, treatment, storage, and so on), 3) funding a recycled water program (more difficult than you might imagine), 4) public involvement (absolutely critical to a successful program), and 5) international topics. This 2004 publication also covers individual state recycled water guidelines and regulations, but because such regulations are constantly being refined and updated, be sure and check with your specific jurisdiction for the latest local rules, regulations, and BMPs.

Water Quality of Recycled Water

Because there are no federal regulations, states typically establish water quality and treatment requirements for recycled water. The most restrictive regulations (i.e., demanding the highest quality) cover recycled water uses with unrestricted public exposure, such as turf irrigation at a public park. Remember that conventional wastewater treatment does little to alter the underlying quality of the influent wastewater—pretty much what you put in, you get out. This means that conventional water quality parameters such as hardness, trace minerals, and total dissolved solids

(TDS) are not used to adjudge the quality of recycled water. Rather, the concept of water quality becomes wrapped up in protecting public health from any potential microbial exposure via the treated wastewater.

The most typically regulated water quality parameters for recycled water are:

- Biochemical oxygen demand (BOD)
- Total suspended solids (TSS)
- Turbidity
- Coliforms

The regulatory limits for these parameters depend on the nature of the application, with again, the more unrestricted uses requiring higher-quality water. State recycled water quality requirements for food crop irrigation are all over the board. Some states require only secondary treatment and some disinfected tertiary, while others completely prohibit use of recycled water on food crops. Again, be sure and check with your specific jurisdiction for individual requirements.

When recycled water is used in industrial applications such as cooling water or boiler feed, it is usually treated again to remove specific constituents such as ammonia or TDS. For example, the West Basin Municipal Water District of California offers five grades of recycled water for local industries:

- Tertiary
- Nitrified (tertiary with ammonia removal)
- Pure reverse osmosis (RO)
- Softened RO (RO plus lime softening)
- Ultrapure RO (double-pass RO)

There is one microbial consideration for industrial recycled water use: if a mist is created (say in a cooling tower application), certain coliform limits have to be met.

Indirect Potable Reuse

Although the most widespread use of recycled water is for irrigation (both turf and agricultural), there are many other ongoing uses, including a fairly controversial one—indirect potable reuse. Indirect potable reuse is recycled water that is *purposely* discharged into either groundwater or surface water that ultimately supplies a public drinking water system. It is a contentious issue with many of the green maze's stakeholders—elected officials, regulators, the media, and activists—weighing in on the subject. Opponents have used terms such as *Frankenwater* or described an indirect potable reuse project as "toilet to tap." Over the past 15 to 20 years, indirect potable reuse projects have witnessed both success and failure.

Clearly, because the resulting water is intended for human consumption, this application involves more complex public health implications than simple nonpotable recycling, and will presumably receive a higher degree of treatment than recycled water used in nonpotable applications.

There are two basic categories of indirect potable reuse—unplanned and planned. Unplanned indirect potable reuse occurs when a wastewater treatment plant discharges effluent to surface water that serves as source water for downstream communities. It's evident that unplanned, indirect potable reuse of wastewater has been practiced for millennia (hence the old adage, "we all live downstream"). There are many locations in the United States where a city discharges wastewater into a river that serves as the water supply for downstream water users. For example, cities in the Colorado River Basin discharge their wastewater to the river, including the greater Las Vegas, Nev., metropolitan area, which releases more than 180 million gallons per day into Lake Mead; several hundred miles later, this water is withdrawn by the Metropolitan Water District of Southern California, treated, and distributed to water districts throughout Southern California.

There are two primary categories of indirect potable reuse: groundwater replenishment (via spreading or direct injection) and reservoir augmentation. Groundwater replenishment may be practiced to:

- Establish saltwater intrusion barriers near coastal aquifers (5 percent of recycled water use in California and Florida)
- Provide further treatment for future use (via soil-based microbial activity)
- Augment potable or nonpotable aquifers (the percentage of recycled water that can be blended is limited by regulations)
- Provide storage for subsequent retrieval and use (as recycled water)
- Control or prevent subsidence

Reservoir augmentation takes advantage of the detention time in large reservoirs to ensure a renewable and reliable raw water source.

Studies addressing the safety of indirect potable reuse have been conducted for decades. And, despite data supporting the process as safe, indirect potable reuse remains a controversial—and hot-button—issue. A project proponent can describe *unplanned* indirect potable reuse interminably, but in many cases the public still objects to *planned* indirect potable. Any utility embarking on a groundwater recharge or surface water augmentation project with recycled water needs to lay a firm foundation of public outreach and information to garner community support.

Graywater

Before leaving the topic of recycled water, let's address a related type of recycling: graywater. Although graywater "recycles" potable water, it is not *recycled water* as defined by the California Water Code (Section 13050)—"water which, as a result of treatment of waste, is suitable for direct beneficial use." That same code section defines beneficial use(s) as "domestic, municipal, agricultural and industrial supply; power generation; recreation; aesthetic enjoyment; navigation; and preservation and enhancement of fish, wildlife, and other aquatic resources or preserves." Conversely, graywater is *untreated* household wastewater that has not come into contact with toilet waste, kitchen sinks, dishwashers, or laundry water from soiled diapers, that is then reused on site.

Graywater regulations are somewhat murky (no pun intended), and vary widely from state to state. Many states require a permit from the local administrative authority (such as the building department or health department), and many of these standards only allow it to be used for subsurface irrigation—surface irrigation via sprinklers is prohibited.

Graywater supporters are passionate about their cause (in fact, you could call them environmental activists as defined in chapter 3). They claim that it is a simple and practical use of a valuable resource. On the topic of regulations, here is Oasis Design's guidance for regulators: "My advice to regulators is simple: Copy Arizona. Don't copy California, the Uniform Plumbing Code, or anyone else. The Arizona law could use a tune-up, the other laws need overhaul or scrapping" (http://oasisdesign.net/greywater/index.htm).

Despite such ardent entreaties for wider use of graywater, I strongly oppose the formal use of graywater for a number of reasons. (Pouring an occasional bucket of dishwashing rinse water on roses is another matter.) First, it constitutes disposal of wastewater to the ground, which in every other instance in states such as California requires a permit from an applicable regulator agency, such as the Regional Water Quality Control Board. Second, most households have difficulty changing the HVAC air filter or water filter in their refrigerator in a timely fashion; improperly operating and maintaining a plumbed graywater system affords an opportunity for significant exposure to pathogens, thereby creating a public health threat. Additionally, an in-home, dual-plumbed system also affords the opportunity for cross-connections between graywater and potable water pipes, which could place the public water supply at risk. Finally, many utilities now produce tertiary-treated recycled water that is protective of public health. Graywater use reduces the volume of water available for safe, publicly beneficial water recycling.

As a water professional dedicated to the protection of public health, I would like to see utilities require sites (even residences) using hard-plumbed graywater systems to be equipped with backflow prevention devices. Most water utilities and municipalities have aggressive and effective backflow prevention requirements for industrial and commercial customers. Similar protective measures should be applied to graywater systems.

However, much like recycled water regulations and standards, graywater regulations continue to evolve at a rapid pace as well. For example, on July 30, 2009, the California Department of Housing and Community Development (HCD) moved to adopt new graywater standards. "These new graywater standards will be a big step toward reducing California's water consumption by providing cost-effective guidelines that will be beneficial to every home throughout the state," according to a press release from HCD Director Lynn L. Jacobs. According to HCD, the new standards include provisions that will:

- Provide a definition for on-site treated graywater and direction to a quality standard
- Encourage the use of new and expanding technology by providing more flexibility through fewer mandatory requirements
- Encourage the installation of inexpensive systems without the cost of design and permits. Such graywater systems:
  - ▲ Must allow users to direct water to either the irrigation field or the sanitary sewer via a diverter valve
  - ▲ Must be designed to prevent ponding and graywater runoff
  - ▲ May be configured to allow graywater to be released aboveground, but the discharge point must be covered by at least 2 inches of mulch, rock, or other material that minimizes human contact
  - ▲ Are prohibited from reusing water used to wash diapers or other soiled garments, which must be directed to the sanitary sewer
  - ▲ Must be configured so they do not discharge into areas where root vegetables are grown

Despite these new standards, I stand by my previously stated opinion that such systems represent a potential hazard to public health.

One final note—an excellent resource, *White Paper on Graywater* (Sheikh 2010) was jointly published by the American Water Works Association, the Water Environment Federation, and the WateReuse Association in early 2010. It provides a thorough summary of graywater issues and outlines both the pro and con arguments for graywater use. It concludes by offering some suggestions for future graywater use policies and research.

CHAPTER 10

OTHER LAWS AND REGULATIONS IMPACTING WATER AND WASTEWATER UTILITIES

It's easy to understand why environmental laws such as the SDWA and the CWA are primary regulatory drivers for water and wastewater utilities. And because utilities produce air emissions and generate hazardous waste, it's not unreasonable to assume further regulation under RCRA and the CAA. However, there are other environmental laws and regulations that might not come to mind so readily, but that can seriously impact water and wastewater utility capital projects and daily operations. Let's take a look at a few of those and how they might twist the green maze a little bit more.

The Endangered Species Act (a.k.a. Bugs and Bunnies)

The term *bugs and bunnies* is often used (in a somewhat derisive tone) by project engineers to describe those activities relating to endangered species investigations when a biological review mandated by NEPA (or a related state environmental review statute such as the California Environmental Quality Act or CEQA) is under way. But a smart project engineer will be well advised to take a broader and more enlightened view: biological issues can impact a project and, as will be discussed, even a state's water supply.

The relevant federal law in the majority of these cases is the 1973 Endangered Species Act (ESA), which is administered by the US Fish and Wildlife Service. There are a host of provisions and requirements under the ESA, among them:

- Listing of species as endangered and threatened
- Establishing critical habitat requiring the identification and protection of all lands, water, and air necessary to recover endangered species
- Prohibiting unauthorized taking, possession, sale, or transport of endangered species

It is an enormously controversial law that has given rise to a wide range of strong opinions as to its value—or harm.

One biological impact review component is to determine if (and how) a proposed project might adversely affect local flora and fauna that are on the Endangered Species List. Qualified biologists are required to conduct surveys to determine the extant natural conditions of the proposed project area and whether any endangered or sensitive species are present. They must then analyze the potential impacts of the project on those flora and fauna. In one recent example, Western Municipal Water District of Riverside, Calif., was required to delay a project because the then-current environmental conditions (a serious lack of rainfall) prevented biologists from conducting a *representative survey* for an endangered butterfly species (the Quino checkerspot butterfly

A snail darter: The fish that stopped a dam. Photograph courtesy of US Fish and Wildlife Service

[*Euphydryas editha quino*]). Following a near-normal winter, the California Department of Fish and Game determined that conditions were appropriate for the butterfly, and a survey was conducted. No evidence of the butterfly was found, and those findings were incorporated into the final CEQA document—a Mitigated Negative Declaration—that allowed the pipeline project to proceed.

A far more famous example of the ESA's impact on a project was the saga of the snail darter (*Percina tanasi*) and the Tennessee Valley Authority's (TVA) Tellico Dam. The Supreme Court ruled in favor of the tiny fish in 1978's *TVA v. Hill* (US 153), a decision that halted construction of TVA's Tellico Dam project. The snail darter was only discovered in 1973, and biologists estimated a modest population of just 10,000 to 15,000 fish—all located in riparian areas near the proposed Tellico Dam. The darter was placed on the recently established Endangered Species List and the waters of the Little Tennessee River identified as critical habitat. A later congressional action exempted the dam project from the ESA, and it was completed in 1979. Interestingly, small populations of the snail darter were both transplanted to, and found in, nearby waters, and its status is now listed as threatened rather than endangered.

A State's Water Supply

Between 2002 and 2007, the Metropolitan Water District of Southern California (the enormous regional water supplier) received annual allocations ranging from a minimum of 1,413,322 acre-feet (2002) to a maximum of 1,801,035 acre-feet (2004). An acre-foot of water is 325,850 gallons, so that's a lot of water. But below-normal precipitation in the northern Sierra Mountains in 2008 ended with record dry conditions, resulting in below-normal snow water content and low runoff projections for the California State Water Project watershed. This may be considered an *environmental drought.*

Several fish species listed as threatened or endangered under the federal and/or California ESAs also have impacted State Water Project operations and limited the flexibility of State Water Project deliveries. Beginning in 2005, separate litigation matters were filed by environmental groups alleging that biological opinions prepared by fish and wildlife agencies inadequately analyzed impacts on protected fish species in the Sacramento–San Joaquin River Delta. On May 25, 2007, a US district court issued a decision finding the US Fish and Wildlife Service's biological opinion for delta smelt to be invalid under the federal ESA (*Natural Resources Defense Council v. Kempthorne et al.*).

This finding was followed on December 14, 2007, by a court order requiring that the State Water Project and Central Valley Project operate according to certain specified criteria until a new biological opinion for the

Oroville Dam with Lake Oroville beyond . . . the lake stores winter and spring runoff, which is released into the Feather River to meet the State Water Project's needs. Photograph courtesy of California Department of Water

delta smelt could be issued. This order resulted in severe cutbacks—up to 30 percent—in the amount of water available for pumping to central and southern California. Thirty percent of the historical State Water Project deliveries is also a lot of water, and this might be termed a *regulatory drought*.

Numerous other court cases and resultant decisions about the fate of the delta smelt have followed the *Kempthorne* decision. All have impacted the available water supply. Recently, the California Department of Fish and Game and the Fish and Game Commission listed the longfin smelt as a candidate species under the California ESA. The longfin smelt is a small fish species that is related to the threatened delta smelt and whose habitat also includes the Sacramento–San Joaquin Delta. Under the California ESA, once a species is granted candidate status, it is entitled to protections until the commission determines whether to list the species as threatened or endangered. Other Sacramento–San Joaquin River Delta species (including salmon) also may be listed as endangered, further impacting State Water Project operations and deliveries. Litigation in these and related cases remains ongoing at the time of this writing (late 2010).

Based on just these two examples, it's clear that water and wastewater professionals must take natural resource issues—the irreverently termed bugs and bunnies—seriously. Mankind shares the world with its other inhabitants, and protecting and safeguarding those inhabitants is an important part of an environmental professional's duties and responsibilities.

Global Warming

Do you believe in global warming? More importantly, do you believe in anthropogenic global warming (AGW)—the concept that mankind is somehow contributing to global warming (or climate change) with CO_2 and other greenhouse gas emissions? Quite honestly, it doesn't matter how you answer those questions, because regulations intended to address AGW already have been promulgated in California and may soon exist as federal law as well.

In 2006, the California legislature passed and Governor Schwarzenegger signed the Global Warming Solutions Act of 2006 (known as AB 32). This law directs the California Air Resources Board (CARB) to begin developing early actions to reduce greenhouse gases to 1990 levels by 2020. The reduction measures to meet the 2020 target are to be adopted by the start of 2011. Among the AB 32 requirements:

- Prepare a scoping plan for achieving the maximum technologically feasible and cost-effective reductions in greenhouse gas emissions from sources or categories of sources of greenhouse gases by 2020.
- Quantify the statewide level of 1990 greenhouse gas emissions, which will serve as the emissions limit to be achieved by 2020.
- Adopt a regulation requiring the mandatory reporting of greenhouse gas emissions.
- Identify and adopt regulations for discrete early actions that could be enforceable on or before January 1, 2010.
- Ensure early voluntary reductions receive appropriate credit in the implementation of AB 32.
- Convene an Environmental Justice Advisory Committee (EJAC) to advise CARB in developing the scoping plan.

- Appoint an Economic and Technology Advancement Advisory Committee (ETAAC) to provide recommendations for technologies, research, and greenhouse gas emission reduction measures.

Again, regardless of your opinion about the validity of AGW theories, the program is inarguably ambitious and will clearly affect all activities in California (not just water and wastewater utilities) on an operational and economic basis. Such actions may be especially problematic in a depressed economy. Interestingly, a voter initiative—Proposition 23—that sought to postpone implementing AB 32 until California's economy recovered and unemployment dropped was defeated (61.6% to 38.4%) by voters in November 2010. Ironically, CARB had released its draft regulations to establish the framework and requirements for California's greenhouse gas cap-and-trade program—some 3,000-odd pages—just five days before the election.

AB 32 is just a state regulation, but it may soon be joined by nationwide regulations; on June 26, 2009, the US House of Representatives passed the American Clean Energy and Security Act of 2009 by a vote of 219 to 212. This legislation would establish 1) a nationwide greenhouse gas cap-and-trade system (to reduce emissions) and 2) complementary measures to help address climate change and build a clean energy economy. As of this writing, the Senate has yet to act on similar legislation.

Do these policies make sense environmentally, economically, socially? As you ponder these questions, keep in mind the following statements from individuals on both side of the climate debate. The disbeliever position was written in July 2009 by Richard S. Lindzen, the Alfred P. Sloan Professor of Atmospheric Sciences at Massachusetts Institute of Technology:

> The notion of a static, unchanging climate is foreign to the history of the earth or any other planet with a fluid envelope. The fact that the developed world went into hysterics over changes in global mean temperature anomaly of a few tenths of a degree will astound future generations. Such hysteria simply represents the scientific illiteracy of much of the public, the susceptibility of the public to the substitution of repetition for truth, and the exploitation of these weaknesses by politicians, environmental promoters, and, after 20 years of media drum beating, many others as well. Climate is always changing. We have had ice ages and warmer periods when alligators were found in Spitzbergen. Ice ages have occurred in a hundred thousand year cycle for the last 700 thousand years, and there have been previous periods that appear to have been warmer than the present despite CO_2

> levels being lower than they are now. More recently, we have had the medieval warm period and the little ice age. During the latter, alpine glaciers advanced to the chagrin of overrun villages. Since the beginning of the 19th century these glaciers have been retreating. Frankly, we don't fully understand either the advance or the retreat.
>
> . . . Wasting resources on symbolically fighting ever present climate change is no substitute for prudence. Nor is the assumption that the earth's climate reached a point of perfection in the middle of the twentieth century a sign of intelligence. (Lindzen 2009)

The other side is represented by Dr. Rajendra Pachauri, Chairman of the Intergovernmental Panel on Climate Change (IPCC), in his opening statement at an August 30, 2010, press conference at the United Nations in New York:

> By overwhelming consensus, the scientific community agrees that climate change is real. Greenhouse gases have increased markedly as a result of human activities and now far exceed pre-industrial values. I think it is worth taking a few minutes to quote from some of these reports. The Muir Russell review from the United Kingdom had this to say: "We did not find any evidence of behavior that might undermine the conclusions of the IPCC assessments." And this from the Dutch Environmental Assessment Agency about the IPCC: "Overall the summary conclusions are considered well-founded and none were found to contain any significant errors." The US Environmental Protection Agency concluded that "climate science is credible, compelling and growing stronger." (IPCC 2010)

Complications

Complicating the AGW issue even further: In December 2009, revelations about possible data manipulation and data suppression by the Climatic Research Unit of the University of East Anglia in Norwich, England, have led some to believe that the "settled science" of AGW may not be so settled after all. Some critics suggest that well-known graphs, such as the "hockey stick," have been altered and/or truncated to emphasize AGW trends and reinforce the posited link of increasing atmospheric CO_2 concentrations and rising global temperatures. Other detractors point to quashed data that illustrated higher temperatures during the 1930s, and faulty analyses that exaggerated the limited warming that did occur during the latter half of the twentieth century.

More importantly, on December 7, 2009, the USEPA issued an Endangerment Finding (USEPA 2009a) under the CAA stating unambiguously that greenhouse gases threaten public health and the environment (USEPA 2009b). The agency further claimed that, yes, the science was settled and "overwhelmingly shows greenhouse gas concentrations at unprecedented levels due to human activity" (USEPA 2009c). This finding is perhaps the first step in USEPA regulating both mobile and stationary CO_2 sources unless Congress regulates them under the Waxman-Markey Bill and its Senate counterpart. Almost immediately, voices such as Patrick J. Michaels (Distinguished Senior Fellow in the School of Public Policy at George Mason University and Senior Fellow of Environmental Studies at the Cato Institute) and Paul C. Knappenberger (a climate researcher at New Hope Environmental Services Inc.) objected to USEPA's conclusions and stated:

> The Proposed Endangerment exemplifies the systematic errors that accumulate at the interface between science and policy when policymakers (in this case, the US Environmental Protection Agency) rely upon previously assembled "compendia" of science, such as the IPCC's AR4 report or the CCSP [US Climate Change Science Program] documents. It is particularly dangerous, in the area of global warming, when an entity takes this approach rather than conducting a new assessment of the scientific literature. (Michaels and Knappenberger 2009)

AWWA and Climate Change

AWWA weighed in on the topic of climate change in a January 17, 2010, Statement of Policy on Public Water Supply Matters that conveyed the Association's concerns regarding this issue:

> The American Water Works Association recognizes that global climate changes will have major impacts on the hydrologic cycle and, therefore, have the potential to significantly affect water supplies. AWWA supports the development of more refined global climate models and tools to better understand and address these impacts at a water utility-relevant scale.
>
> AWWA encourages water utilities to: support regional climate change modeling; respond proactively to anticipated climate changes; develop adaptation and mitigation responses for a range of predicted climate change impacts; and reduce energy consumption and greenhouse gas emissions as reasonably feasible. (AWWA 2010)

Water Supply Availability Requirements

Growing populations and the increasing scarcity of reliable, high-quality water supplies have led several states to pass laws linking new residential and industrial development with water supply availability. Some common features of these laws are requirements to

- Evaluate current water supply conditions noting potential constraints and shortfalls
- Describe future water supply projects, their status, and likelihood of completion (no "paper water")
- Assess and quantify the water demands of the proposed development project
- Describe conservation or mitigation measures that could be implemented to reduce demand
- Analyze the supply and demand data
- Project whether adequate water supplies are/will be available for both the existing customer base and the proposed development project.

These water supply availability reports are intended to answer a simple question: is there a sufficient water supply to serve the proposed project over the statutory planning horizon? A big catch is that the planning horizon can stretch into decades. We'll examine two recent pieces of water supply availability legislation: California's SB 610/SB 221 and Colorado's HB 1141.

California SB 610/SB 221

Senate Bills 610 (SB 610) and 221 (SB 221) amended California state law on January 1, 2002. As the State Water Resources Control Board SB 610 guidance relates:

> [T]hese bills were intended to improve the link between information on water supply availability and certain land use decisions made by cities and counties. SB 610 and SB 221 are companion measures which seek to promote more collaborative planning between local water suppliers and cities and counties. Both statutes require detailed information regarding water availability to be provided to the city and county decision-makers prior to approval of specified large development projects. Both statutes also require this detailed information be included in the administrative record that serves as the evidentiary basis for an approval action by the city or county on such projects. (California Department of Water Resources 2003)

The "detailed information" stipulated is packaged in a report known as a Water Supply Assessment (WSA). SB 610 requires the preparation of a WSA for certain projects that meet any of the following criteria:

- A residential development of more than 500 units
- A business/shopping center with more than 1,000 employees or 500,000 square feet of floor space
- A commercial office building with more than 1,000 employees or 250,000 square feet of floor space
- A hotel/motel with more than 500 rooms
- An industrial/manufacturing/processing plant or industrial park with more than 1,000 employees, encompassing more than 650,000 square feet of floor space, or occupying more than 40 acres of land
- A mixed-use development project or other project with water demands equal to 500 equivalent dwelling units (EDUs)
- A project that would demand an amount of water equivalent to, or greater than, the amount of water required by a 500-dwelling-unit project, or
- For a supplier with 5,000 or fewer connection, a project that will increase the number of connections or total demand by more than 10 percent

The WSA must analyze whether the total projected water supplies available to the water provider during normal, single-dry, and multiple-dry years over the *next 20-year period* is sufficient to meet the projected water demand associated with the project, in addition to existing and planned future uses. Once the WSA is adopted by the water provider's board (or council), it must be forwarded to the lead agency and incorporated into the planning document being prepared for the project. Then, the lead agency (*not* the water utility) must determine, based on the entire record, whether projected water supplies will be sufficient to satisfy the demands of the project, in addition to existing and planned future uses.

Given the ongoing environmental and regulatory droughts in California, issuing WSAs has become problematic for many water utilities. What will snowfall (and thus spring runoff) be in the Sierras in 2018? Will the state have the political will to construct a parallel water conveyance (or *peripheral canal*) to convey that water south? What will the region's population be in 2022? A water purveyor can predict supply and demand—but how accurate will those predictions be?

Some water utilities have placed temporary moratoriums on issuing WSAs, stating that the water supply situation is too uncertain to allow accurate prognostication some 20 years into the future. Others exhibited cautious optimism and concluded that potential local water supply projects would be adequate to support the assurances conveyed by the WSA.

Colorado HB 1141

Even though the state of Colorado contains the headwaters of many rivers, including the mighty Colorado, the South Platte, and the Arkansas among others, it is still a relatively arid state. Water scarcity was never more apparent than during the extraordinarily severe summer 2002 drought. Many water utilities along the Front Range instituted mandatory water conservation measures, while some newer cities (with the most junior water rights) completely prohibited all outdoor watering. So it's not surprising that there are concerns about ensuring adequate water supplies for future growth in that state. (See chapter 12 on water law for an overview of water rights around the country.)

The Colorado legislature attempted to address these water supply concerns by passing House Bill 08-1141 or HB 1141, *Concerning Sufficient Water Supplies for Land Use Approval*. Effective May 29, 2008, HB 1141 requires that developers of projects consisting of more than 50 single-family equivalents (that is, 50 single-family homes or a development with water demands equal to or exceeding 50 single-family homes) provide local governments with information regarding the adequacy of proposed water supplies for the project.

Like its California counterparts, HB 1141 contains a 20-year planning horizon and its statutory language requires project proponents to:

- Estimate ultimate (that is, at build out) water supply requirements for the proposed development
- Describe the physical source of water supply that will be used to supply the proposed development
- Estimate annual water yield of that water supply under various hydrologic conditions
- Describe water conservation measures that will be implemented by the development
- Describe water demand management measures that may be implemented by the development

If the development is served by a local water utility, the law requires that water purveyor to submit a report detailing essentially this same information. After analyzing all the submitted information, much like the California law, it's up to the local government to determine whether there is an adequate water supply to serve the proposed development.

In the coming years, it's likely that water supplies will become even more stretched; it's equally likely that the contentious issue of development and the water available for such development won't disappear either. Thus, it's *highly* likely that the use of regulatory mechanisms such as SB 610/SB 221 and HB 1141 will become only become more widely employed by planning agencies across the country, and not just in the arid West.

Bioterrorism Preparedness and Response Act

The September 11, 2001, terrorist attacks on this country profoundly shook the nation; on September 12, citizens awoke to an uneasy awareness of an entirely new world of threats. Drinking water utilities weren't immune from this awareness as utility managers pondered the potential for malicious anthropogenic (or human-caused) attacks on their infrastructure and ultimately on the safety of the drinking water they provided to customers.

The 2001 terrorist attacks prompted Congress to pass the Bioterrorism Preparedness and Response Act of 2002, which required community drinking water systems serving populations of more than 3,300 persons to conduct assessments of their susceptibility to terrorist attack or other intentional acts. These *vulnerability assessments* (VAs) identified critical drinking water infrastructure (including information technology [IT] and supervisory control and data acquisition or SCADA networks) and their potential vulnerabilities, and provided recommended improvement measures for facilities, programs, and practices.

The Bioterrorism Act also required utilities to certify and submit their VAs to USEPA. Because these often lengthy documents contained a huge amount of extremely sensitive information about a utility, USEPA was directed to rigorously safeguard the VAs and their contents. Throughout 2003 and 2004, thousands of community water systems across the country assessed their infrastructure and operations, and took a hard look at the anthropogenic threats that could adversely impact them, including terrorists, vandals, thieves, and insider threats. The VAs drove a host of corrective actions to physical security systems (fences, locks, and alarms), IT/SCADA systems, and operational procedures (access control and staff identification). It was a tremendous effort and resulted in a higher degree of security for this critical national infrastructure sector. I'm proud that in consort with numerous other water professionals, I conducted or managed more than 50 vulnerability assessments for water utilities across the country.

CHAPTER 11

STATE ENVIRONMENTAL REGULATIONS

The Federal/State Regulatory Nexus

Recall that for most federal environmental laws, it was always the intent of Congress for the individual states to take the lead managing and enforcing the requirements established at the federal level. States could promulgate their own environmental regulations if the requirements were *at least as stringent as federal standards.* Some states, like California, promulgated entirely new regulations—drinking water, hazardous waste—that in structure and context bore no resemblance to federal regulations. Other states simply adopted the federal regulations by reference; thus their regulatory language is identical to the various 40 *CFR* citations.

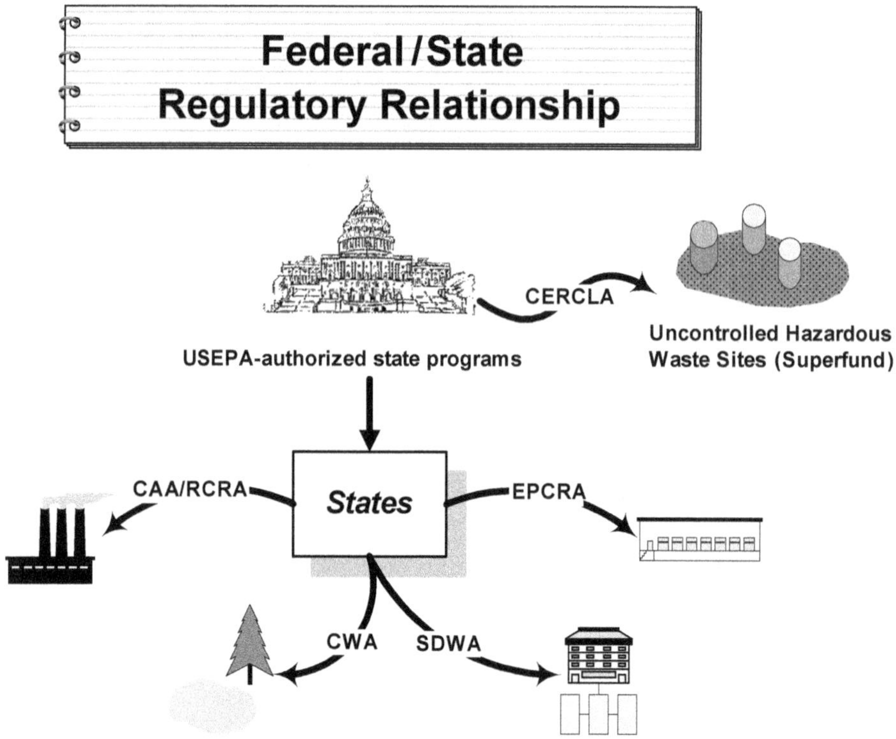

However, it's not hard to imagine that there are difficulties in this federal–state relationship. As Golden Gate University School of Law Professor Clifford L. Rechtschaffen and Florida State University College of Law Professor David L. Markell write:

> Most federal environmental statutes currently operate under a framework known as "cooperative federalism," in which states receive primary authority to implement the requirements of federal laws, under EPA supervision. Over the years, states have gradually assumed "primacy" in more and more areas; they now have authority to implement approximately three-quarters of the major federal environmental programs that can be delegated to them, and carry out in the neighborhood of eighty percent of the enforcement actions under the federal statutes. While states have long resented aspects of EPA oversight, some states traditionally relied on the presence of the federal "gorilla in the closet" to motivate compliance by regulated entities. But in the 1990s, tensions between EPA and the states grew considerably. States began vigorously pressing for increased autonomy and flexibility from the strictures of EPA oversight. (Rechtschaffen and Markell 2002)

Regulatory Enforcement: USEPA v. the States

One of the most contentious federal–state intersections lies in the enforcement arena. USEPA delegates enforcement responsibility to states, but the states must satisfy the agency that they can effectively manage the regulatory program in a manner that meets (or exceeds) all federal requirements. The hitch is, USEPA retains supervisory authority and can rescind delegation at any time. Although frequently threatened, USEPA has never (to my knowledge) totally revoked a state's authority to run a regulatory program. Such an action would require the federal agency to assume the responsibility of administering that program in a state—an expensive, staff-heavy, and time-consuming process at best.

Several observations and conclusions contained in a 1995 Government Accounting Office (GAO) report, *EPA and the States: Environmental Challenges Require a Better Working Relationship*, are probably still valid today, especially with the ongoing state budget problems across the nation:

> Most states authorized to manage federal environmental programs have at times been unable to meet some of the requirements for implementing these programs. Many states have had difficulty performing key functions, such as monitoring environmental quality, setting standards, issuing permits, and enforcing compliance. Consequently, states have become increasingly reluctant to accept the additional responsibilities associated with recent environmental laws.
>
> EPA and state officials uniformly acknowledged that resource limitations are a major cause of these problems: Federal funding has not kept pace with new environmental requirements, and the states have been unable to make up the difference. Many EPA and state officials contacted by GAO linked this resource gap to the "unfunded mandates" debate that has gained widespread attention in the Congress and elsewhere in recent years. The resource shortage, however, has been exacerbated because EPA has sometimes required states to apply scarce resources to national priorities at the expense of some of their own environmental concerns.
>
> Also affecting the EPA/state relationship have been states' concerns that EPA (1) is inconsistent in its oversight across regions, (2) sometimes micromanages state programs, (3) does not provide sufficient technical support for increasingly complex state program requirements, and (4) often does not adequately consult states before making key decisions affecting them. (US GAO 1995)

USEPA can take enforcement action into its own hands if it believes that an authorized state has not adequately addressed a violation, a process known as *overfiling*. One of the more well-known instances of this occurred during the 1990s in *Harmon Industries, Inc. v. Browner* (Carole Browner was the USEPA administrator at the time). Harmon Industries was a Missouri-based printed circuit board manufacturer that had been illegally disposing of solvents on their site for many years. When management became aware of this practice, they voluntarily contacted the Missouri Department of Natural Resources (MDNR) and reported the violation. Missouri, and the MNDR, was (and remains) an RCRA-authorized state responsible for regulating hazardous waste management (including transportation, storage, treatment, and disposal).

During the investigation and subsequent negotiations, MDNR and Harmon developed a plan to remediate the site, and Harmon requested that MDNR not impose any monetary penalties. MDNR notified USEPA, which 1) recommended an $800,000 penalty and 2) initiated a separate enforcement action seeking an additional $2.3 million in penalties! While the USEPA action was still pending, Harmon and MDNR settled in a state court with no monetary penalty levied against Harmon. Despite this settlement, a USEPA administrative law judge assessed a $586,000 penalty against Harmon. Naturally, Harmon appealed, maintaining that the case had already been settled in state court.

The Environmental Appeals Board (there is an entirely separate administrative judicial structure for environmental laws; note previous mention of the administrative law judge) affirmed and Harmon appealed again, this time in federal trial court, which overturned the penalty. The court determined that USEPA's decision to impose civil penalties violated RCRA, opining that the state program operated "in lieu" of the federal program. This meant that federal action (by USEPA) was ruled out if the state had been delegated enforcement authority (which it had). USEPA tried, unsuccessfully, to argue that it retained authority to independently pursue enforcement as part of its overfiling authority.

Here's what the court essentially said to USEPA: look, you can initiate enforcement *only* if a state decides not to, but an authorized state *did* take action; you just didn't agree with the outcome. On further appeal, the Eighth Circuit Court (a federal appellate court) affirmed the trial court's decision and ruled that the concept of overfiling was unconstitutional, reasoning that RCRA's statutory structure and principles of federalism (allowing the states to enforce federal laws) bars USEPA from enforcing Missouri law. The appellate court further stated that if USEPA was dissatisfied with how Missouri was enforcing RCRA, its only recourse was to completely revoke its regulatory authorization and take over all RCRA activities in the state. USEPA elected not to do so.

The results in *Harmon* were both praised and criticized. In addition to reaffirming the concept of federalism, supporters of the outcome stated that it reinforced the notion that state-level agencies were better positioned to enforce environmental regulations and protect the environment. Critics, on the other hand, questioned this logic by asking, what's to prevent a state from enticing industry to locate within their borders with the promise (via a wink and a nod) of lax or favorable enforcement of environmental regulations?

Adjudicating the principle of overfiling continues. In a more recent case (2002's *United States v. Power Engineering Co.*), the Tenth Circuit Court (another federal appellate court) reaffirmed USEPA's capacity to overfile in an enforcement action. The ultimate decision on the practice of overfiling must wait until a case reaches the US Supreme Court.

One Notable Exception

One notable exception to state enforcement of a federal regulation is CERCLA or Superfund law. USEPA retains primary authority for all Superfund activities throughout the country in the Office of Solid Waste and Emergency Response (OSWER). Superfund sites are proposed and identified on the National Priorities List or NPL (many actions related to site listings on the NPL can be found in the daily *Federal Register*). Within OSWER, the Office of Emergency Management conducts short-term responses while the Office of Superfund Remediation and Technology Innovation manages the long-term Superfund response program. OSWER also oversees the federal Brownfields Program, which is intended to remediate (for reuse and redevelopment) impacted industrial (usually urban) sites. USEPA does involve other federal, state, tribal, and local agencies in determining the selected Superfund response and remediation, but they function in an ancillary or support role.

State Regulatory Structures

Every state and territory has its own unique environmental laws, regulatory agency structure, and compendium of applicable regulations. Simply reciting state by state regulatory citations and references wouldn't be terribly productive—or useful. As the federal regulatory maze shifts and changes, so too do the smaller mazes in the states and territories. The list would essentially be worthless in six months to a year. So, what to do?

Well, once again, the magic that is the Internet is the answer. There are two excellent online resources that can be accessed for an up-to-date listing of state regulatory agencies and their associated regulations. One is hosted by the USEPA; the other, EnvCAP, is operated by the National Center for Manufacturing Sciences, which is "an industry consortium specializing in cross-sector collaboration" that runs the Web site for the benefit of the regulated community. Interestingly, EnvCAP receives support from USEPA's Office of Enforcement and Compliance.

The URL of USEPA's state-specific regulatory Web site is http://www.epa.gov/lawsregs/states/index.html, and the Web page looks something like this:

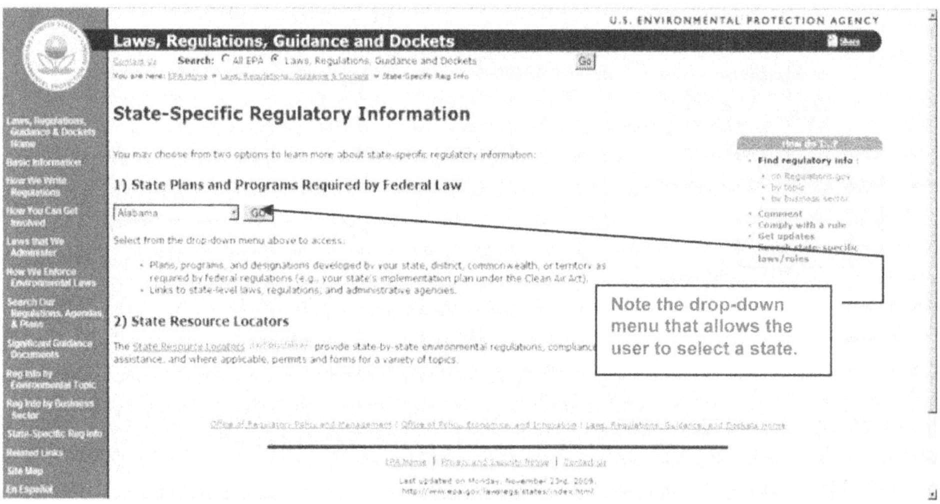

Use the drop-down menu to access, for example, the individual page for the state of Illinois, which in turn provides a list of additional links:

Illinois

On this page, you can access information about the plans and programs your state has put in place in response to federal requirements. Links will direct you to the Electronic Code of Federal Regulations (e-CFR) where you can find the sections pertinent to your state. In addition, we provide links to your state's regulatory information.

State Plans/Programs under Federal Regulations

- Air
- Lead-Based Paint
- Waste
- Water

State Regulatory Information

These are all links to informational URLs.

If you are looking for more general information about EPA's regulations, please visit our How We Write Regulations or Basic Information pages.

State Plans/Programs under Federal Regulations

Air

Your state's air quality planning designation areas:

- Air Quality Control Regions Interstate and Intrastate
- Air Quality Control Region Numbers (see Appendix)
- Section 107 Attainment Status

State Plans

- Approved Implementation Plan
- Approved Plans for Designated Facilities and Pollutants

Lead-Based Paint

- State Authorization Application (September 1, 1999 version)
- State Program Approval—Interim
- State Authorization Application (January 11, 2002 version)
- State Program Approval—Final

Waste

- Approved Hazardous Waste Management Program

Water

- Underground Injection Control Program

State Regulatory Information

- State of Illinois Home Page Link to EPA's External Link Disclaimer
- Legislature Link to EPA's External Link Disclaimer
- Administrative Agencies Link to EPA's External Link Disclaimer

EnvCAP's URL is http://www.envcap.org/, and its associated subsite for state regulations is found at http://www.envcap.org/statetools/. It's a great service and an extremely useful tool for environmental professionals.

CHAPTER 12

WATER LAW (THE SHORT, NON-ATTORNEY VERSION)

Before we go any further, let me state right here that I am *not* an attorney; I am especially not a water law attorney. I have worked with many environmental and water law attorneys during my career and I have an enormous degree of respect for them and their abilities. There's an old bit of folk wisdom, "A man who acts as his own attorney has a fool for a client." Well, that goes double if an engineer acts as his or her own environmental or water law attorney.

This relatively short chapter is intended to (briefly) introduce water and wastewater environmental professionals to the *extraordinarily complex* topic of water law. Water law deals with the ownership, control, and use of water as a resource. It answers the basic questions: Who owns the water? Who controls it? Who has rights to use it? How can it be used?

Mark Twain, a man wise in the ways of western water.
Photograph courtesy of Library of Congress

The first thing to remember is that, in most jurisdictions, the state itself "owns" the natural or public water within its boundaries. Water rights, which are discussed in the following sections, address the diversion and *use* of that water. In the United States, there are byzantine legal systems in nearly every state for allocating water rights. How those rights are assigned varies by region; and these varying systems exist for both historical and geographic reasons.

So let's dive into the subject of water law (and yes, that was a pun).

Let's say that you've just purchased a lovely plot of riverfront land and have decided to set up a drinking water utility. Can you just drop the intake hose of your ACME 5000 drinking water treatment plant in the river and start sucking water into it? The answer to that question depends on where that lovely riverfront land is located. If it's in the eastern part of the country, the answer is probably yes, you can use the water. However, if it's in the western part of the United States, well, there's a saying in the West (which is apocryphally attributed to Mark Twain): "Whiskey's for drinking, water is for fighting over."

Riparian Rights

The doctrine of riparian water rights is rooted in English common law. The eastern states (all those east of Texas, except Mississippi) follow what is termed *riparian doctrine*, which permits anyone whose land fronts on a body of water to use water from it. These eastern states were the first settled by Europeans (and therefore most influenced by English law) and have the most available water.

A riparian right entitles the landowner to use a correlative (or proportional) share of the water flowing past his or her property. There is no right to significantly divert streamflow, store water, or adversely impact water quality because such actions would impair the rights of downstream users. Additionally, uses must be for legal and beneficial purposes.

Riparian water rights cannot be lost through nonuse and last indefinitely (which differs significantly from prior appropriation).

Prior Appropriation (a.k.a. "First in Time, First in Right")

Most western states, which are drier than their East Coast counterparts, generally follow the prior appropriation doctrine, which assigns a water right to whomever first puts water to beneficial use, hence the phrase, "first in time, first in right." This is a key concept, because unlike with riparian rights, the appropriator must use—or demonstrate intent to use—the water, and apply it to beneficial use. Typically, individual states define what constitutes beneficial use, and such uses may include domestic water supply, industrial and commercial uses, agriculture, recreation, and support of various natural habitats.

> There's another saying that's been attributed to (among others) Glen Sanders or Marc Reisner (author of *Cadillac Desert*): ***In the western United States, water flows uphill toward money.***

Colorado water law is generally looked to as an authority by other western states that follow the prior appropriation doctrine. In Colorado, there are water courts and water judges, who are district judges appointed by the Colorado Supreme Court. They have jurisdiction in determining water rights as well as the use and administration of water.

Water law in the western United States is defined by state constitutions, statutes, and case law. Each state exhibits variations upon the basic principles of the prior appropriation doctrine. Texas and the states directly north of it, the West Coast states, and Mississippi have a mixture of systems. Hawaii uses a form of riparian rights, and Alaska uses appropriation-based rights.

And those three paragraphs more or less wrap up all you'll ever need to know about prior appropriation.

Hybrid Systems of Water Law

Many states, including California, look to water laws that are hybrid systems with elements of both riparian and prior appropriation rights. For example, in California all waters are considered property of the state and thus a water right is a use right. Riparian rights are present, but appropriative rights can be obtained by putting water to beneficial use. Owners of overlying land may extract groundwater for beneficial use; however, some basins are adjudicated and court decrees govern groundwater rights and extraction. One of the premier texts on the complex topic of water law in the Golden State is *California Water Law*, which was coauthored by Arthur L. Littleworth, a partner in the Riverside, Calif., office of Best, Best & Krieger.

Interstate and International Water Compacts and Agreements

There's a funny thing about rivers—they don't necessarily flow within the artificial political boundaries of just one state. Sometimes they mark state boundaries, such as the Mississippi River between Missouri and Illinois; other rivers flow across state lines and through adjacent states, eventually becoming a boundary before crossing a national border, such as the Colorado River. What if water rights differ in the states the river traverses; which water rights apply in these cases? To further complicate the matter, what if the river flows between two countries? What happens then?

Interstate Water Compacts

Well, in the case of interstate disputes, states have turned to the US Constitution and the compact clause, Article I, Section 10:

> *No State shall, without the Consent of Congress*, lay any duty of Tonnage, keep Troops, or Ships of War in time of Peace, *enter into any Agreement or Compact with another State*, or with a foreign Power, or engage in War, unless actually invaded, or in such imminent Danger as will not admit of delay.

Thus, states may, with congressional consent, enter into agreements or *compacts* to solve common problems, which in this case involve water rights. The authority of the federal government to intervene in these interstate water wars was affirmed by the Supreme Court's 1907 decision in *Kansas v. Colorado* (206 US 46). There are a number of famous (or infamous, if you prefer) interstate water compacts:

- *Colorado River Compact.* Signed in 1922 by representatives of the seven states within the 250,000-square-mile Colorado River basin (Upper Division [Colorado, New Mexico, Utah, and Wyoming] and the Lower Division [Nevada, Arizona, and California]), the compact governs water rights allocations along the river's 1,450-mile length. Commonly known as the Law of the River, this compact has been controversial since its inception with ongoing disputes that frequently require federal involvement by the

Department of the Interior to cajole (and occasionally threaten) the various parties into revised agreements and allocations. Even with the compact, the states have frequently resorted to litigation such as in the 1963 Supreme Court decision *Arizona v. California*, which eventually led to the 1968 Central Arizona Project. There are a host of other, ancillary agreements between various parties (for example, California Seven Party Agreement of 1931, Colorado River Storage Project of 1956, Colorado River Basin Project Act of 1968) that continue to tweak and refine the Law of the River.

COLORADO RIVER BASIN
Divisions of upper and lower basins

This map provides some idea of the size of the Colorado River Basin and the complexity of the Law of the River

- *Arkansas River Compact.* Ostensibly settling a nearly 50-year dispute between Colorado and Kansas, this 1949 agreement apportioned Arkansas River water deliveries and created the Arkansas River Compact Administration as the supervising agency. However, much like the Colorado River Compact, there has been a history of ongoing litigation between the two parties with years of judgments, rulings, and damage settlements. Best, Best & Krieger's Arthur Littleworth (mentioned earlier as author of *California Water Law*) acted as a Special Master (a court-appointed representative who hears a case involving difficult or specialized issues) in many of these actions.

International Water Compacts

The Colorado River runs through seven US states, but eventually outfalls into salt marshes at the head of the Gulf of California in Mexico. The Mexican Water Treaty of 1944 guaranteed delivery of 1.5 million acre-feet of the river's annual flow to Mexico. A later codicil, Minute 242 of the US–Mexico International Boundary and Water Commission of 1973, mandated reduced salinity of Mexico's allotted Colorado River water. The 1973 agreement resulted in the Bureau of Reclamation's Yuma Desalting Plant, which was constructed to treat highly saline agricultural return flows from the Wellton-Mohawk Irrigation and Drainage District. Although completed in 1992, actual operations have been minimal due to hydrologic conditions on the Colorado.

To the north, there are a number of compacts and agreements between the United States and Canada, including the 1909 Boundary Waters Treaty, which was intended to resolve water disputes (regarding both quantity and quality) along the US–Canadian border, and the 1972 Great Lakes Water Quality Agreement, which pledged both countries to restore and maintain the chemical, physical, and biological integrity of the Great Lakes Basin ecosystem. It further reaffirmed the rights and obligations of Canada and the United States under the earlier 1909 treaty.

CHAPTER 13

WATER AND WASTEWATER UTILITIES ORDINANCES

We've seen how federal and state environmental laws and regulations impact construction projects and the daily operations of water and wastewater utilities. Do utilities have similar internal environmental laws and regulations that govern their users? They do, and they're usually called ordinances, or as *Webster's Ninth New Collegiate Dictionary* defines it, "a law set forth by a governmental authority, specifically a municipal regulation."

But not all water or wastewater utilities are part of a municipal (or city) government. Water and wastewater utilities come in many sizes—and many shapes.

- Some, such as the Golden (Colo.) Public Works Department, which oversees the Water and Wastewater Divisions, are indeed part of a city government and report (ultimately) to the mayor and/or city council.

- Others, like Western Municipal Water District in Riverside, Calif., are what are known as *special districts*, which are units of local government established by the residents of an area to provide some service not provided by the county or city (it doesn't have to be a water or wastewater utility; there are park and recreation districts, flood control districts, vector control districts, and so on). Western was established by voters in 1954 for the express purpose of bringing imported water (supplied by the Metropolitan Water District of Southern California) to western Riverside County. As an independent district, Western operates under a locally elected, independent board of directors composed of five people who represent different geographical areas or divisions.
- Still other water (and wastewater) utilities, such as the 100+-year-old Aqua America, Inc. (formerly the Philadelphia Suburban Corporation/Philadelphia Suburban Water Company), are for-profit enterprises with stockholders (equity owners); these are commonly known as *investor-owned utilities* or IOUs. Depending on the state it operates in, certain aspects of an IOU's finances (particularly rates and charges) may be regulated by state entities such as a public utility commission or department of public utilities. However, remember that drinking water standards are regulated by the SDWA, irrespective of a utility's ownership or governance.

So, how do utilities operate? Typically, the governing board enacts ordinances, which are simply the utility's rules and regulations. They have the force of law but are limited to the purpose of the district. In other words, although it might be handy, a water utility can't stipulate and enforce a speed limit on the street fronting its building.

A water utility's ordinances may address a number of issues, but in general they focus on:

- General use regulations
- Rights and responsibilities of the utility and the customer
- Physical connections/meter installations
- Backflow prevention
- Service requirements and agreements
- Service classifications
- Residential/agricultural/commercial/industrial
- Water use and conservation
- Administrative issues including billing practices, late charges, and rights of appeal before the board

- Rates and charges
- Connection fees
- Monthly service fees (meter fees)
- Commodity charges

One item on this list is water use and conservation. and many utilities include conservation practices and guidelines in their general ordinance. However, with water shortages nearly a constant in some western states, and intermittent droughts in other areas such as Texas and the Southeast, some water utilities have recently enacted explicit water use efficiency and/or conservation ordinances.

For example, on May 26, 2009, the Western Municipal Water District's Board of Directors adopted a new, stand-alone ordinance establishing a Water Conservation and Supply Shortage Program. On what basis did the board pass this ordinance? They relied on the California Constitution, as noted in the ordinance's preamble:

> California Constitution article X, section 2, and California Water Code section 100 provide that because of conditions prevailing in the State of California, it is the declared policy of the State that the general welfare requires that the water resources of the State shall be put to beneficial use to the fullest extent of which they are capable, the waste or unreasonable use of water shall be prevented, and the conservation of such waters is to be exercised with a view to the reasonable and beneficial use thereof in the interest of the people and the public welfare.

Western published public notices announcing to its customers that a hearing would be held regarding this ordinance, and then held that public hearing to allow members of the public to comment (pro and con) on the proposed ordinance. The ordinance outlines various shortage stages and appropriate water efficiency use/conservation measures, which range from simple common sense practices that should always be followed, to a complete prohibition on outdoor irrigation during periods of severe shortages or emergencies. Importantly, the ordinance also includes a series of escalating enforcement measures, up to and including termination of service.

Water utilities that also provide recycled water service generally enact a dedicated recycled water use ordinance. They can cover many of the same topics as potable water ordinances, but may also include requirements for:

- On-site facilities (e.g., purple pipe, hose bibs, signs)
- System testing and cross-connection prevention

- Operational requirements
- Use restrictions
- Operator training

Of course, wastewater utilities and businesses at the other end of the pipe have their own set of ordinances. In addition to addressing the business aspects of providing wastewater services (just as with potable water service), they may stipulate:

- General prohibitions and limitations on discharges (usually tied to the POTW's NPDES permit requirements)
- Pretreatment requirements (from the CWA, see chapter 6)
- Wastewater discharge permits (industrial and commercial customers)
- Miscellaneous control provisions such as FOG (fats, oil, and grease) program
- Enforcement process for pretreatment program violations

CHAPTER 14

ENVIRONMENTAL REGULATIONS AND THE PUBLIC

Public Participation

One of the hallmark features of modern (post-1970) environmental legislation is its emphasis on public participation—during rulemakings, enforcement actions, and project reviews. The USEPA itself has a Public Involvement Policy (USEPA 2003), which was updated in 2003 (see http://www.epa.gov/publicinvolvement/public/index.htm). USEPA states that the 1981 policy was updated because of

- Changing needs of the public
- New statutes and regulations
- New and expanded public participation techniques

- New options for public access to information and involvement through the Internet
- USEPA's emphasis on ensuring compliance
- Increased use of partnerships and technical assistance
- Increased capacity of states, tribes, and local governments to carry out delegated programs

The policy applies to all USEPA activities and programs and provides guidance and direction to USEPA officials on ways to involve the public in its regulatory and program decisions. The revised policy "reaffirms EPA's commitment to early and meaningful public involvement" (USEPA 2003).

Other nongovernmental entities also espouse and support the concept of public participation. One such example is the American Water Works Association, which has a published a Public Involvement and Customer Communication Policy Statement (emphasis added):

> AWWA recognizes the critical importance and multiple benefits of clear and timely *communication with customers, meaningful involvement by community members and stakeholders, and proactive, frank information-sharing regarding water quality and service.*
>
> Water providers have a key role in support of public well-being. They manage a crucial natural resource that is essential to people's health, community economies, and the fabric of social stability. Water quality and reliability no longer can be considered as solely technical; they are health, lifestyle, economic, social, and political issues.
>
> AWWA supports integration of *communication and public involvement as core elements of utility management. The credibility and reputation of utilities and the water profession as a whole are supported by open and ongoing communication that establishes relationships and creates a framework for understanding, trust, and cooperation in times of crisis.*
>
> *Communication must be a dialogue with customers, community members, and stakeholders.* This dialogue should involve a broad range of topics and issues of importance to consumers and stakeholders, including water quality, reliability, conservation, infrastructure investment, and costs for service. Messages should be consistent, factual, and tailored to the receptivity and interest of specific audiences, including sensitive and special populations.
>
> Opportunities for input and involvement are essential to public understanding and acceptance of utility programs and

> projects. The format and depth of involvement will vary according to individual suppliers, communities, and issues. Opportunities for involvement must, however, be meaningful, inclusive, and clearly linked to the decision-making process.
>
> *AWWA advocates ongoing communication and involvement as beneficial to the publics and communities served by water utilities.* Applied with honesty, openness, and receptivity, these critical aspects of utility planning and operations broaden understanding and support for the growing challenges of providing safe, reliable drinking water while advancing awareness of the value of water. (AWWA 2007)

To the average customer, water utilities deliver water for drinking or washing or irrigating the lawn; wastewater utilities take care of the yucky stuff. But that's not the real purpose or mission of water and wastewater utilities; they are all about *public health*. This point cannot be emphasized strongly enough. In my opinion, all the other services provided by water and wastewater utilities are just ancillary. Their primary mission is *public health*. Until mankind conquered water supplies and sanitation, millions of people died from waterborne illnesses; sadly, millions still die worldwide from preventable causes.

Because water and wastewater services are such an essential component of public health, it is critical that *the public* be involved in the many aspects of delivering water and providing wastewater services. AWWA gets it right: "ongoing communication and involvement" is not only beneficial, but crucial for a utility's (and the industry's) ultimate success.

Numerous other regulatory schemes also require public involvement, most notably during the NEPA (or state-level programs such as CEQA) review process. Public involvement is intended to ensure that the public has input in the decision-making process. The process mandates that public concerns about environmental issues and the potential impacts of a proposed project are addressed prior to project approval.

Public Communication

One topic related to public involvement is the concept of public communications. Here too, many environmental regulations promote—or require—effective and meaningful public communication. This section will focus on the public communications *requirements* of the SDWA. There are two primary components of SDWA-related public communication: 1) routine informational communication and 2) notifying the public about problems with their drinking water system.

Cover of Western Municipal Water District's 2010 CCR

Routine Communication

One of the most important aspects of the 1996 SDWA Amendments was the Consumer Confidence Report Rule. This rule requires drinking water utilities to publish and distribute an annual Consumer Confidence Report (CCR) to customers that contains information about their drinking water: its source, how it is treated, water quality information (MCLs), information on detected contaminants, and if applicable, compliance information. The CCR supplants but does not substitute for timely notification (discussed next) of anomalies or problems. Many utilities also use this document as a yearly opportunity to publicize capital projects or new conservation programs.

Emergency Public Notification

The general public takes for granted—and should, given the professionalism of US water utilities—that the water streaming out of their tap is safe to drink. When, for whatever reason, it is not safe or healthy to consume, it is incumbent upon drinking water utilities to notify their customers in a timely and appropriate manner. This is known as public notification.

Public notification has always been part of the SDWA regulatory mix. Minor changes in the nature and manner of how communications are delivered were promulgated in 2000, primarily to require faster notification to the public. Thus the primary goal of public notification is to *immediately alert consumers if there is a serious problem with their drinking water that may pose a risk to public health.* But there are also requirements to notify customers of other, less serious violations or variances from required treatment technologies.

Depending on the nature of the violation or circumstances requiring public notification, a tier rating is assigned (from 1 to 3, with 1 being the most serious), which mandates public notification (and the means/method of notification) within 24 hours, 30 days, or one year (respectively). Examples of Tier 1 violations requiring immediate (24-hour) notification are:

- Fecal coliform MCL violation (perhaps requiring a boil-water order)
- Nitrate/nitrite/combined nitrate and nitrite MCL violation
- Waterborne disease outbreak or other waterborne emergency (i.e., serious contamination possibly resulting in a do not use order)

Some Tier 2 violations require notification within 30 days:

- Other MCL or treatment technique violations not included in Tier 1

Tier 3 violations require notification within one year (usually via the annual CCR):

- All other monitoring or testing procedure violations not already requiring a Tier 1 or Tier 2 notice
- Operating under a treatment variance or exemption

USEPA released a revised and updated version of its excellent resource *Revised Public Notification Handbook* (Revised Document: 816-R-09-013, March 2010), which is available for downloading at http://www.epa.gov/safewater/publicnotification/pdfs/Revised-Public-Notification-Handbook-CWS.pdf.

The agency even provides Microsoft Word® templates for public notices to ensure that the appropriate statutory language and warnings are provided to the public (see http://www.epa.gov/safewater/publicnotification/compliancehelp_templates.html from which the list on page 110 is obtained).

Microsoft Word® Public Notice Templates Available from USEPA

Tier 1

- Nitrate Notice
- Spanish Nitrate Notice
- Fecal Coliform or *E. coli* Notice
- Spanish Fecal Coliform or *E. coli* Notice
- Waterborne Disease Outbreak Notice
- Notice for Turbidity Single Exceedance as Tier 1
- Chlorine Dioxide Maximum Residual Disinfectant Level (MRDL) (Tier 1) Notice
- Tier 1 Problem Corrected Notice

Tier 2

- Unresolved Total Coliform Notice
- Resolved Total Coliform Notice
- Chemical or Radiological MCLs Notice
- Fluoride MCL Notice
- SWTR Failure to Filter Notice
- SWTR Turbidity Exceedance Notice
- SWTR Disinfection Treatment Notices
- Lead and Copper Rule (LCR) Failure to Install Corrosion Control

Tier 3

- Monitoring Violations Annual Notice
- Fluoride Secondary Maximum Contaminant Level (SMCL) Notice

For noncommunity systems

- Nitrate Notice
- Fecal Coliform or *E. coli* Notice
- Unresolved Total Coliform Notice
- Monitoring Violation Notice

Translated Phrases (in Spanish, French, Vietnamese, Chinese, and Korean)

- This report contains important information
- Don't drink the water
- Boil your water before using
- Don't use this water for infant formula

Risk Communication

The USEPA uses the definition of *risk assessment* found in a 1983 publication by the National Research Council: "a process in which information is analyzed to determine if an environmental hazard might cause harm to exposed persons and ecosystems" (National Research Council 1983). It is a study or investigation. Although there are prescribed methodologies and practices, it is not an exact science; uncertainly and variability are inherent in risk assessments.

On the other hand, *risk management* is the process of evaluating the results of a risk assessment, and then determining the degree of protection (via regulations or some other control mechanism) required to protect public health and the environment at acceptable levels. Risk management or environmental decision making is often (oh, let's just say it—*always*) a controversial process. There are competing interests, conflicting priorities, and "uncertain science" (you can't study something forever—you finally have to make a decision and issue a report). Too lax a standard means potentially endangering public health or the environment, but a standard that is too strict may provide only marginal public benefits at excessive costs.

Good risk communication?

Risk communications is the process of explaining to a (usually) scientifically (and mathematically) challenged populace the whys and wherefores of a decision or regulation or policy. For example, how do you explain that a risk level of 1×10^{-6} is much better than one of 1×10^{-4} when first, negative exponents are a toughie and second, most folks want $0 \times 10^{-\text{whatever}}$ risk! It's pretty self-evident that it is a difficult and daunting task.

Even experts in the field have difficulty simply defining the term *risk*; one common definition for the noun *risk* is "possibility of injury or loss: PERIL." As a verb it means to expose something to hazard or danger. Depending on how it's used, you'll find subtly different definitions in the fields of law, finance, occupational safety and health, and medicine.

Then there is the issue of absolute risk and comparative risk. Comparing risks with different dimensions is tricky and can often backfire.

Why? Because humans *perceive* different risks differently. In a 1984 text, *Acceptable Risk*, Baruch Fischhoff and his colleagues (Fischhoff et al. 1984) observed that people's perceptions of the magnitude of risk are influenced by factors other than numerical data (so let's drop the 1×10^{-6} stuff right now).

- Risks perceived to be voluntary are more accepted than risks perceived to be imposed.
- Risks perceived to be under an individual's control are more accepted than risks perceived to be controlled by others.
- Risks perceived to have clear benefits are more accepted than risks perceived to have little or no benefit.
- Risks perceived to be fairly distributed are more accepted than risks perceived to be unfairly distributed.
- Risks perceived to be natural are more accepted than risks perceived to be man-made.
- Risks perceived to be statistical are more accepted than risks perceived to be catastrophic.
- Risks perceived to be generated by a trusted source are more accepted than risks perceived to be generated by an untrusted source.
- Risks perceived to be familiar are more accepted than risks perceived to be exotic.
- Risks perceived to affect adults are more accepted than risks perceived to affect children.

Hours could be spent discussing each of these points, but the question remains: how do public health professionals (and that most definitely includes water and wastewater professionals) communicate risks of their projects, actions, and policies? In his contribution (chapter 8.9) to a later text, *Oxford Textbook of Public Health* (Detels et al. 2002), Fischhoff posited the following (I have helpfully included my simplistic interpretations of Fischhoff's guidance):

- Complete mental models, by bridging the gaps between expert and lay mental models. (I believe the good professor means—keep it simple, stupid.)
- Ensure appropriate confidence in beliefs. The most dangerous beliefs are those held with too great or too little confidence. (A high degree of public trust is mandatory; maintain it at all costs . . . oh, and be humble . . . but not too humble.)
- Provide information in order of its expected impact on decisions. (Don't delve into the minutiae before explaining the important stuff.)

Once the appropriate information is selected, it must be communicated. Fischhoff recommends clear, concise verbal (or written) with adjunct aids, such as highlighting, advanced organizers (showing what to expect), and summaries. He concludes with, "[E]ffective risk communication is essential to managing risks in socially acceptable ways. Without it, individuals are denied the best chances of making sound choices, before, during, and after problems arise" (Detels et al. 2002).

Environmental Advocacy

> **advocacy** \ad'vo-ca-cy\ *n.* [OF *advocatie*, LL *advocatia*. See ADVOCATE.] The act of pleading for or supporting; work of advocating; intercession.
>
> **advocate** \ad'vo-cate\ *n.* [OE *avocat*, *avocet*, OF *avocat*, fr. L *advocatus*, one summoned or called to another; properly the p. p. of *advocare* to call to, call to one's aid; *ad* + *vocare* to call. See ADVOWEE, AVOWEE, VOCAL.]
>
> 1. One who pleads the cause of another.
>
> 2. One who defends, vindicates, or espouses any cause by argument; a pleader; as, an advocate of free trade, an advocate of truth.

Recall that one stakeholder group in the drinking water/wastewater regulatory universe was identified as "environmental activists." This is a broad category and, as noted, could include a single individual concerned with a particular contaminant in drinking water, a local group lobbying to use a drinking water reservoir for recreation, or an ad hoc neighborhood group opposing the construction of a new storage tank.

But for the sake of argument, let's define an *environmental advocate* as a member of a more formal organization that continually works to influence public policy in the environmental arena: "one who pleads the cause of another." Although an advocate can be a group supporting (advocating) the construction of a new desalination plant, an industrial facility, or a particular environmental ruling, I think most environmental advocates are those groups and their members who are typically concerned with social, economic, political, and cultural issues relating to environmental regulation and natural resources. Some example organizations are:

- Environmental Defense Fund
- Environmental Law Institute

- Friends of the Earth
- Greenpeace USA
- League of Conservation Voters
- Natural Resources Defense Council
- Public Citizen
- Rainforest Action Network
- Sierra Club

Many of these organizations consist not only of volunteer (and dues-paying) members, but paid professional staffs with natural science or legal backgrounds. Their staff attorneys often take advantage of the citizen suit provisions of environmental regulations (notably under the CWA and CAA) to bring suit against industries and utilities for alleged statutory violations when enforcement is not pursued by regulatory agencies.

So, are these environmental advocates friends of the earth or just environmental nuisances? Perhaps they're a little of both. Often these groups are staffed by individuals with "strong commitments to social justice and environmental policy reform" as described by EnvironmentalPrograms.net (EnvironmentalPrograms.net 2010). But these well-intentioned individuals frequently bring a strong bias as evidenced by this statement: "we've focused our resources where we can make the biggest impact in . . . corridors of power and boardrooms of commerce, where environmental issues are often overlooked or ignored" (Minnesota Center for Environmental Advocacy). Representatives of this group probably enter any negotiations already prejudging the people on the other side of the table as uncaring despoilers of the environment.

Despite being public health emissaries, water and wastewater professionals should understand that there are many environmental advocacy groups that oppose much-needed water supply projects and initiatives. Some examples of controversial California projects that are opposed (at least in part) by some environmental advocacy groups are the Oceanside/Poseidon Desalination Plant, Alternative Delta Supply Conveyance (a.k.a. the Peripheral Canal), and the Cadiz Inc.'s storage and groundwater mining project.

Another recent example is the inflammatory report released in December 2009 by the Environmental Working Group (EWG), intimating that drinking water was tainted by "315 pollutants in the tap water Americans drink" (EWG 2010a). This 501(c)(3) organization further fanned the flames, stating, "More than half of the chemicals detected are not subject to health or safety regulations and can legally be present in any amount. The federal government does have health guidelines for others, but 49 of these contaminants have been found in one place or another at levels above those guidelines, polluting the tap water for 53.6 million

Americans. The government has not set a single new drinking water standard since 2001" (EWG 2010a). The last sentence presumably refers to the new arsenic MCL of 10 µg/L, which indeed was promulgated in 2001, effective in 2002, and required compliance in 2006.

Do those statements seem like a fair, reasoned, and objective view of the water industry? More importantly, did this report contribute to public health or did it (as stated on EWG's Web site) "bring[s] to light unsettling facts that you have a right to know. It [EWG] shames and shakes up polluters and their lobbyists" (EWG 2010b). Water professionals who ignore these groups, their causes, and their power do so at their own peril.

Before leaving the topic of environmental advocacy, remember that an advocate can represent many points of view—including those of industry. And much like the Natural Resources Defense Council and others, industry advocates are not beyond using the American legal system to achieve their goals. The notice below was published in the July 30, 2009, *Federal Register* and concerns a CAA action (note the groups that filed the lawsuit: the Business Coalition for Clean Air [BCCA] Appeal Group, Texas Association of Business, and Texas Oil and Gas Association):

> In accordance with section 113(g) of the Clean Air Act, as amended ("CAA" or "Act"), 42 U.S.C. 7413(g), notice is hereby given of a proposed consent degree and proposed settlement agreement to address a lawsuit filed by the Business Coalition for Clean Air (BCCA) Appeal Group, Texas Association of Business, and Texas Oil and Gas Association in the United States District Court for the Northern District of Texas, Dallas Division in BCCA Appeal Group, et al. v. EPA, No. 3-08CV1491-G (N.D. Tex.). Plaintiffs filed a complaint in this action on August 25, 2008, which alleges that EPA has failed to perform nondiscretionary duties pursuant to section 110(k)(1)(B) and (k)(2) of the Federal CAA, 42 U.S.C. 7410(k)(1)(B) and (k)(2), to take final action on numerous Texas State Implementation Plan ("SIP") air quality revisions by the statutory deadline. The SIP revisions are related to Texas New Source Review (NSR) air permitting and banking and trading rules. The proposed consent decree and proposed settlement agreement, if finalized, will collectively resolve plaintiffs' claims in this matter. A schedule for acting on the SIP revisions is attached to the proposed consent decree and proposed settlement agreement.

CHAPTER 15

THE ROLE OF LABORATORIES—PARTNERS IN ENVIRONMENTAL COMPLIANCE

How do water utilities know that they comply with the SDWA? How do they prove their compliance to their state oversight authority? How do wastewater utilities know that their treatment plant effluent complies with their NPDES permit? How does the public have confidence that high levels of pollutants aren't being discharged into a river?

Today's modern water and wastewater treatment plants use a variety of online instrumentation to monitor performance: pH meters, turbidity meters, flowmeters, and so on. But simply relying on these instruments will not suffice to ensure compliance—a more reliable and accurate measure is needed. This is the role of the environmental laboratory.

Today's modern environmental laboratory doesn't bear much resemblance to this 1919 version. Photograph courtesy of Otis Historical Archives of the National Museum of Health & Medicine http://www.nmhm.washingtondc.museum/

Online instruments are accurate, but today's state-of-the-art analytical laboratories are far more accurate—often by orders of magnitude (see "The Vanishing Zero" below). Because water and wastewater utilities rely so heavily on environmental laboratories, they are truly partners not only in environmental compliance, but public health as well.

Many larger water and wastewater utilities own and operate their own laboratories; some of these only perform limited analysis, while others have greater capabilities. One large regional utility, the Passaic Valley Water Commission (New Jersey), operates an analytical laboratory capable of performing virtually any organic, inorganic, or microbial analysis. However, most medium and small utilities use the services of a contract environmental laboratory such as century-old E.S. Babcock & Sons Laboratories, located in Riverside, Calif., which analyzes drinking water, wastewater, groundwater, stormwater, soils, and hazardous materials.

Every drinking water permit, every NPDES permit, and many air permits require periodic chemical or microbial analyses to ensure compliance. Depending on the requirement, some media are sampled daily, others quarterly, still others annually. Not only are there detailed analytical methods, but there are stringent sampling protocols as well.

But whether a utility uses its own or employs a contract laboratory, the laboratory must be certified and adhere to a strict set of operational, analytical, and reporting protocols. For example, Babcock Laboratories is certified by the California Department of Public Health (and has been since 1928) for the analysis of drinking water for public health protection, one of the first in the state to receive this certification.

Laboratory and Custody Protocols

Holding public trust is vital; the public and regulatory agencies must be confident that the analytical results demonstrating compliance are real and representative. This means that the entire sequence of events, from taking the sample, transporting it to the laboratory, analyzing the sample, and reporting the results must be verifiable and documented. The integrity of the entire sampling/analytical process must be ensured if regulatory and public confidence is to be maintained.

Analytical laboratories engage in extensive employee education, not just in physically how to conduct an analysis, but also in the ethical standards expected during that process. E.S. Babcock publishes and updates an internal document that describes the framework by which the laboratory establishes and maintains its high standards. Such a program is required by the National Environmental Laboratory Accreditation Conference for a laboratory to maintain its accreditation status.

Babcock also publishes a code of ethics to assure its customers of the reliability of their analytical processes and procedures. Among the topics covered are:

- Data integrity
- Fraudulent actions
- Inappropriate practices
- Examples of unethical behavior
- Pressures that could lead to unethical behavior
  - Examples of external pressures (clients):
    - "The result is wrong—it is above my MCL. I will be fined."
    - "It's only a little bit out—couldn't you round it?"
    - "I'll go somewhere else if my results are too high."
  - Examples of internal pressures (employees/management):
    - "If the QC fails, I will have to repeat it."
    - "Get it out NOW!"

    - "No time—I have personal plans."
- Response to potential ethical breaches
- Disciplinary actions

The same degree of rigor that an E.S. Babcock applies to conducting laboratory analyses must extend to the upstream process of sampling. Required sampling, transportation, and chain-of-custody documentation procedures must be followed; there is no "dry lab-ing" in the water and wastewater industry. Integrity is paramount—water and wastewater professionals must maintain the confidence of the public and regulatory agencies.

The Vanishing Zero

Before leaving the topic of laboratories, let's look at the increasing analytical prowess of today's state-of-the-art analytical laboratories and what that means to the water and wastewater community.

Some 50-odd years ago, Ivory Soap used the advertising slogan "99 and 44/100ths percent pure!" Wow, that's pretty pure, right? Well, let's look at it from a different perspective: 99 and 44/100ths percent pure means that 56/100ths *aren't* pure and are composed of . . . contaminants? So 0.56 percent is something else: 1 percent is 1/100, so 0.56 percent is .56/100 or 56/10,000 or 56,000/1,000,000 or 56,000 parts per million (ppm) (or, in water-speak, milligrams per liter [mg/L]). The MCL for nitrate (as N) is 10 ppm! Doesn't seem that pure today, does it?

Actually, one-half of one percent is a small quantity, and for many things (like soap), measuring impurities at those levels of precision is just fine. But what about other substances, like trace organics, pesticides, or heavy metals in drinking water? Do we want to measure at the percent level? Probably not.

The advances in analytical chemistry during the twentieth century are nothing short of phenomenal. At the turn of the twentieth century, using wet chemistry (reagents, colorimetric indicators, flame spectroscopy, and mechanical balances), a good analytical chemist could achieve part-per-million accuracy. As the century progressed, new analytical instruments—gas chromatographs, liquid chromatographs, and mass spectrometers—allowed laboratories to measure ever smaller and smaller concentrations of stuff in air, water, and soil. Water that once was thought relatively pure and free from impurities was discovered to be not so pure after all.

These advances were paralleled by advances in epidemiology and the understanding that trace amounts of substances could be harmful to human health, especially over long periods of time. Science was able to see ever smaller concentrations of contaminants, and medical professionals were able to say that those even smaller concentrations posed potential human health or environmental risks.

By the latter half of the twentieth century, most commercial laboratories had long passed the part-per-million barrier and were capable (depending on the analyte) of detecting compounds in the part-per-billion range (micrograms per liter [μg/L]). This is the concept of the Vanishing Zero. Increasing analytical prowess reduced the concept of zero contamination of "purity" from water that simply appeared clear to the naked eye to parts per billion of some contaminant. But science still wasn't done. The 1990s and early 2000s witnessed further analytical progress and the ability to analyze (some) substances at the part-per-trillion (ppt) (nanograms per liter or ng/L) level! Parts per trillion: let's look at that.

$$\frac{1}{100} = 1\ \%$$

$$\frac{1}{1{,}000{,}000} = \text{1 part per million (ppm or mg/L)}$$

$$\frac{1}{1{,}000{,}000{,}000} = \text{1 part per billion (ppb or μg/L)}$$

$$\frac{1}{1{,}000{,}000{,}000{,}000} = \text{1 part per trillion (ppt or ng/L)}$$

...

..

.

ZERO

By anybody's standards, those are a lot of zeros. But what do those zeros mean? More importantly, what does that lonely little 1 mean on top of all those zeros? It's 1 of something, but does that 1 of something pose a risk to human health or the environment? Just because we *can* measure something and thus have redefined the definition of *pure,* does it mean we are now at risk?

This is a dilemma that is currently facing water and wastewater professionals in the arena of endocrine-disrupting chemicals (EDCs, or the more ubiquitously termed *emerging contaminants*), which will be discussed in the next chapter. But for now, consider the following statements made by Shane Snyder (formerly of the Southern Nevada Water Authority, now an associate professor of aquatic science at Harvard University School of Public Health) in his brilliant testimony before Congress in April 2008. He was discussing his recent study on EDCs, the preliminary results of which had been widely reported in an AP story the month before.

> A more central point about our studies is that the few pharmaceuticals we did detect in US drinking waters occurred at unfathomably low concentrations. To illustrate that point, consider this: If our study had been constrained by the ability to find these compounds at parts-per-billion levels instead of delving into the parts-per-trillion range, none of them—not a single one—would have been found.
>
> This raises a critical question. Are we going to make decisions based upon our ability to find contaminants, or based upon protection of public health? I am not a policy-maker; I am a scientist. However, I can tell you with absolute certainty that, if we regulate contaminants based upon detection rather than health effects, we are embarking on a futile journey without end. The reason is simple: Decades ago, we could only detect contaminants at parts-per-million levels. Years ago, we advanced to parts per billion. We are now able to detect compounds at the parts-per-trillion level, and are breaching the parts-per-quadrillion boundary in some cases. The fact that we can detect trace contaminants does not alone imply risk.
>
> With regard to removing these compounds through treatment, my team has tested the effectiveness of a diverse array of water treatment technologies on removal of pharmaceutical compounds, and to be certain, some technologies are more effective than others. However, the pinnacle question is whether the use of these treatment technologies is warranted to protect public health, because there are environmental and societal costs associated with using them.
>
> In an age where we are concerned about greenhouse gas emissions and minimizing our nation's energy demands, is it wise to dictate energy-intensive water treatment systems when there is no evidence of public health benefits? Additionally, there is a looming crisis related to aging water infrastructure that will require a vast financial investment by utilities. Should that be set aside so they can chase down the last nanogram of a compound? (Snyder 2008)

Later in his testimony, Snyder asked Congress to let USEPA and the scientific community do their jobs under the auspices of the SDWA and its new Unregulated Contaminant Monitoring Rule and the Contaminant Candidate List to identify the nature, extent, and scope of these EDCs and then let epidemiologists and health professionals do their jobs and determine if these minute quantities pose a risk to human health and the environment. In other words, don't make a hasty and ill-informed decision and establish policies, regulations, and controls before the industry even knows if a problem really exists.

CHAPTER 16

MORE MAZES: TOMORROW'S REGULATIONS

Here we are at Chapter 16, and it's been a long and winding journey through the green maze of water and wastewater utilities regulations—and remember, if you're going to make a career in this industry, you'll never get out of that maze. But what's on tap for tomorrow? Next year? Twenty years from now? Is the green maze going to get longer, shorter, more complex? Well, if I could accurately answer all those questions, I'd be making a fine living in a number of Las Vegas sports books.

But looking back at the past 35+ years of modern environmental legislation, I'd say it's unlikely that the maze is going to get shorter or less complex. The better bet is just the opposite: more regulations with ever-increasing complexity. So with that basis in mind, let's take a look at what might be around the next corner of the green maze.

In the near term, the federal government provides a twice-a-year regulatory look ahead when it publishes the Unified Agenda (UA), which provides information about regulations that the government is considering or reviewing. You can access recent and past UAs at http://www.gpoaccess.gov/ua/index.html.

There have been no legislative rumblings about a new set of amendments for the Clean Air Act. However, the agency continues to promulgate Maximum Achievable Control Technology standards under National Emission Standards for Hazardous Air Pollutants for various industrial source categories and area sources in consonance with the 1990 Amendments. See http://www.epa.gov/ttn/atw/mactfnlalph.html for a complete list. Should USEPA eventually regulate greenhouse gases including CO_2, I suspect that much of its regulatory emphasis will be in that arena.

On the Clean Water Act front, two Supreme Court decisions, 2001's *Solid Waste Agency of Northern Cook County v. United States Army Corps of Engineers et al.* (No. 99–1178) and the 2006 case *Rapanos et ux., et al. v. United States* (No. 04–1034), caused considerable consternation (and confusion) within the USEPA, the Corps of Engineers, and the regulated community alike. Both of these cases essentially center on the CWA definition of *waters of the United States*—as in "what are they?" and "how are they to be regulated?"

The controversy and confusion surrounding these cases forced the agency to develop a separate Web page on its Internet site (http://www.epa.gov/owow/wetlands/guidance/CWAwaters.html) just to provide information, updates, and guidance regarding the CWA and "waters of the United States." Not surprisingly, following the June 2007 issuance of post-*Rapanos* guidance, the USEPA received a staggering 66,047 public comments! In mid-2009, the Democrat-chaired Senate Environment and Public Works Committee reported out a bill that would restore to waterways environmental protections that some believe were eroded by the two referenced Supreme Court decisions. Other senators oppose the bill, with Senator Mike Crapo (R-Idaho) stating in a June 18, 2009, press release, "This bill threatens the current Clean Water Act statute and would allow for government regulation of virtually all interstate and intrastate waters and their tributaries, including rivers, intermittent streams, mudflats, sandflats, prairie potholes, wet meadows, playa lakes, natural ponds, and others." This topic and other recent news headlines regarding CWA shortcomings and increased pollutant discharges may lead to a major amending process for the CWA.

The Safe Drinking Water Act seems to be proceeding along apace under the regulatory structure established by the 1996 Amendments. From information sources such as news releases, announcements, and conferences, we know that USEPA is considering revisions to the Total Coliform Rule and the Lead and Copper Rule. Of course the Unregulated Contaminant Monitoring Rule (UCMR) and the Contaminant Candidate List result in an ongoing process of identifying and evaluating new contaminants that could be subject to future regulatory rulemakings. You can always check out new, proposed, and potential SDWA regulations at http://www.epa.gov/drink/standardsriskmanagement.cfm.

The most recent CCL—number three to be exact—was published in the October 8, 2009, *Federal Register.* The final list contained 116 contaminants scheduled for regulatory evaluation, and differed from the draft list by including one antibiotic and nine hormones. No new contaminants were regulated from the CCL1 and CCL2 lists, and some believe that there is political pressure to regulate *something* from list number three. Here's the CCL3 list in its entirety:

Chemical Contaminants

1,1,1,2-Tetrachloroethane
1,1-Dichloroethane
1,2,3 Trichloropropane
1,3-Butadiene
1,3-Dinitrobenzene
1,4-Dioxane
17alpha-estradiol
1-Butanol
2-Methoxyethanol
2-Propen-1-ol
3-Hydroxycarbofuran
4,4'-Methylenedianiline
Acephate
Acetaldehyde
Acetamide
Acetochlor
Acetochlor ethanesulfonic acid
Acetochlor oxanilic acid
Acrolein
Alachlor ethanesulfonic acid (ESA)
Alachlor oxanilic acid
alpha-hexachlorocyclohexane
Aniline
Bensulide
Benzyl chloride
Butylated hydroxyanisole
Captan
Chlorate
Chloromethane (methyl chloride)
Clethodim
Cobalt
Cumene hydroperoxide
Cyanotoxins
Dicrotophos
Dimethipin
Dimethoate
Disulfoton
Diuron
Equilenin
Equilin
Erythromycin
Estradiol (17-beta estradiol)
Estriol
Estrone
Ethinyl Estradiol (17-alpha ethynyl estradiol)
Ethoprop
Ethylene glycol
Ethylene oxide
Ethylene thiourea
Fenamiphos
Formaldehyde
Germanium
Halon 1011 (bromochloromethane)
HCFC-22
Hexane
Hydrazine
Mestranol
Methamidophos
Methanol
Methyl bromide (bromomethane)
Methyl tert-butyl ether
Metolachlor
Metolachlor ethanesulfonic acid

Metolachlor oxanilic acid
Molinate
Molybdenum
Nitrobenzene
Nitroglycerin
N-methyl-2-pyrrolidone
N-nitrosodiethylamine (NDEA)
N-nitrosodimethylamine (NDMA)
N-nitroso-di-n-propylamine (NDPA)
N-nitrosodiphenylamine
N-nitrosopyrrolidine (NPYR)
Norethindrone (19-norethisterone)
n-Propylbenzene
o-Toluidine
Oxirane, methyl-
Oxydemeton-methyl
Oxyfluorfen
Perchlorate
Perfluorooctane sulfonic acid
Perfluorooctanoic acid
Permethrin
Profenofos
Quinoline
RDX (Hexahydro-1,3,5-trinitro-1,3,5-triazine)
sec-Butylbenzene
Strontium
Tebuconazole
Tebufenozide
Tellurium
Terbufos
Terbufos sulfone
Thiodicarb
Thiophanate-methyl
Toluene diisocyanate
Tribufos
Triethylamine
Triphenyltin hydroxide
Urethane
Vanadium
Vinclozolin
Ziram

Microbial Contaminants

Microbial Contaminant	Information
Adenovirus	Virus most commonly causing respiratory illness, and occasionally gastrointestinal illness
Caliciviruses	Virus (includes Norovirus) causing mild self-limiting gastrointestinal illness
Campylobacter jejuni	Bacterium causing mild self-limiting gastrointestinal illness
Enterovirus	Group of viruses including polioviruses, coxsackieviruses, and echoviruses that can cause mild respiratory illness
Escherichia coli (O157:H7)	Toxin-producing bacterium causing gastrointestinal illness and kidney failure
Helicobacter pylori	Bacterium sometimes found in the environment capable of colonizing human gut that can cause ulcers and cancer
Hepatitis A virus	Virus that causes a liver disease and jaundice

Microbial Contaminants (continued)

Microbial Contaminant	Information
Legionella pneumophila	Bacterium found in the environment including hot water systems causing lung diseases when inhaled
Mycobacterium avium	Bacterium causing lung infection in those with underlying lung disease and disseminated infection in the severely immunocompromised
Naegleria fowleri	Protozoan parasite found in shallow, warm surface water and groundwater causing primary amoebic meningoencephalitis
Salmonella enterica	Bacterium causing mild self-limiting gastrointestinal illness
Shigella sonnei	Bacterium causing mild self-limiting gastrointestinal illness and bloody diarrhea

Emerging Contaminants

One of the hottest topics on both the drinking water and wastewater regulatory fronts is *emerging contaminants*. No, these aren't some new type of exotic chemical that's been cooked up in a mad scientist's laboratory and released on an unsuspecting public. In fact, some of them (endocrine disruptors) have probably been around as long as man has been excreting waste products—we just haven't been able to detect these ubiquitous substances.

NDMA

One emerging contaminant that is a likely candidate for regulation (and found on CCL3) is NDMA or N-nitrosodimethylamine, one of a broader class of chemicals known as nitrogenous disinfection by-products (or N-DBPs). It is used in liquid rocket fuels and other industrial applications. Recent research suggests that it is also a by-product of drinking water treatment including chloramination and anion exchange. NDMA's potential health impacts (as a suspected carcinogen) and frequency of occurrence (as determined by data from the second Unregulated Contaminant Monitoring Rule [UCMR]) place this contaminant (and its sister nitrosamines) squarely in the sights of SDWA regulations. Notably, NDMA detection falls under the realm of the vanishing zero in that it is typically found at concentrations in the nanogram-per-liter (ng/L) or part-per-trillion range. The California Department of Public Health lists a 1×10^{-6} risk level for NDMA at the astoundingly low concentration of 3 ng/L; for those of you more familiar with fractions, that's

$$\frac{3}{1{,}000{,}000{,}000{,}000}.$$

Endocrine Disrupting Compounds

Endocrine disrupting compounds (EDCs) are known by a number of monikers, including *pharmaceuticals and personal care products* (PPCPs) and *emerging constituents*; but for the purposes of this section, they're all going to be lumped together (more or less) into the category of EDCs (it's a shorter acronym). Subject to much general press, including a series of front-page Associated Press (AP) news articles in March 2008, these chemicals have been detected in trace amounts in surface water, drinking water, and wastewater effluent sampling conducted in both Europe and the United States, which is why both drinking water and wastewater professionals need to be concerned with them.

What are EDCs? Simply stated, they are chemicals (both natural and synthetic) that can potentially interfere with the proper functioning of the endocrine system. The endocrine system is a collection of glands (such as the hypothalamus, pituitary, or thyroid glands) that secrete different hormones that regulate such bodily functions as growth, development, tissue function, and metabolism. The presence of low concentrations of these chemicals in surface waters has led to sexual dimorphism in fish, which was widely reported in the AP stories. Direct causal linkage of these substances to potential human health interactions is scanty, but research continues.

These chemicals can be introduced into the environment in several ways, including:

- Flushing unused medications down the toilet or sink
- Rinsing personal hygiene and household cleaning products down the drain
- Excreting unabsorbed medications into the sewage system
- Farm animals excreting veterinary drugs, including hormones and antibiotics, into fields where they run off into lakes and streams
- Improper industrial/commercial wastewater disposal methods

Thus the presence of EDCs in surface water presents a challenge for both drinking water and wastewater utilities.

As we saw in chapter 15 in connection with The Vanishing Zero, recent advances in analytical technologies mean that today, industry has the ability to detect more substances, at lower levels, than ever before. These compounds are being found at levels far lower than where drinking water standards are typically set. But the fact that a substance is detectable in drinking water does not mean the substance is necessarily harmful to humans or the environment. As Shane Snyder testified in April 2008, "To date, no peer reviewed published research has found ill effects on humans from pharmaceuticals in the environment at the trace levels we have seen in drinking water."

As the science of these compounds continue, Snyder provided some near-term recommendations:

> So what should we do? A couple of things make sense. This issue does highlight the need to better protect America's sources of drinking water from various sources of contamination. And clearly there is a pressing need for additional research on this issue. As a scientist, I recommend we focus on research related to health effects from trace pharmaceuticals with a lesser emphasis on occurrence, in order to determine whether there is in fact a problem to solve. The critical question we must address is not "Do they exist?" but rather, "At what concentration are these compounds harmful to human health?" Only then can we make intelligent, rational decisions that protect the health of this country's municipal water customers.

1. USEPA should work with states, water and wastewater utilities, and the agricultural community to minimize contamination of source waters by pharmaceutical products as well as other contaminants.

2. We urge support for proper pharmaceutical disposal programs to reduce the flushing of pharmaceutical products into sewage systems to the greatest degree possible, while recognizing that this addresses only a small part of the problem. Although more research would be needed to accurately characterize this issue, we believe it is likely that more pharmaceuticals end up in the environment after passing through humans than after flushing unused products. However, some unused pharmaceutical products are undeniably flushed into waste streams, contributing to the

> problem but also offering an opportunity to make reductions in the pollutant loading through a "pollution prevention" approach. We urge support for pharmacy "take back" programs that make doing the right thing obvious and convenient for consumers. (Snyder 2008)

Perchlorate

Perchlorate (ClO_4^-) is both a naturally occurring and synthetic inorganic chemical used as the primary ingredient in solid rocket fuel and to a lesser extent in missiles, fireworks, explosives, and air-bag inflators. Highly soluble in water, perchlorate has been detected as a contaminant in groundwater or surface water in 20 states. Human exposure to perchlorate can affect the thyroid gland and cause thyroid tumors. Research continues to determine the potential health risks caused from ingesting low levels of perchlorate in drinking water and to develop an effective treatment. Perchlorate cannot be removed from water by conventional filtration, sedimentation, or air stripping, thus leaving ion exchange, biological treatment, or tight membranes as possible treatment technologies.

Although USEPA issued a preliminary determination notice in October 2008 *not* to regulate perchlorate under the SDWA, continuing public concern and the receipt of numerous stakeholder comments have led the agency to reevaluate this decision. USEPA published an August 2009 Supplemental Request for Comments seeking input on potential perchlorate health impacts, interestingly, segregated by age group (that is, does perchlorate affect different age cohorts differently?); this is a regulatory tack regarding drinking water contaminants that has never been taken before by USEPA. Does this mean perchlorate will soon be regulated under the SDWA? It's hard to say; the American Water Works Association reiterated its earlier stance on the substance stating that it should not be regulated, as it (a new rule) does not present "a meaningful opportunity for risk reduction" (Roberson 2009). But pressure from other stakeholders may lead to eventual regulation by USEPA or individual states.

Pathogens

Many microbiological pathogens have only recently been recognized as being potentially harmful to human health. Under the CCL provisions of the 1996 Amendments, several of these are being investigated for possible inclusion as primary drinking water contaminants, including a dozen on the CCL3 list. Among the more notable CCL3 contenders are:

- *Caliciviruses.* This family includes noroviruses, which are frequently the cause of gastroenteritis (think stomach flu) disease outbreaks on cruise ships.

- *Escherichia coli* (O157:H7). A particularly virulent enterohemorrhagic strain of this ubiquitous bacteria, it was responsible for at least seven deaths (with thousands sickened) during a 2000 outbreak in the small Canadian town of Walkerton, Ontario. This incident resulted in an extensive revamping of Canada's approach to regulating drinking water utilities.
- *Legionella pneumophila*. This gram-negative bacterium is the culprit behind Legionnaire's Disease; it is frequently detected in institutional hot water systems and is a common causal agent of nosocomial infections.
- *Naegleria fowleri*. Infections from this parasitic amoeba characteristically found in warm, shallow freshwater ponds are rare, but unfortunately, usually fatal.

INTERNET RESOURCES

- US Environmental Protection Agency *http://www.epa.gov/*
  - Office of Air and Radiation *http://www.epa.gov/air/*
  - Office of Groundwater and Drinking Water *http://www.epa.gov/ogwdw/index.html*
  - Office of Solid Waste and Emergency Response—RCRA and CERCLA regulations and emergency response *http://www.epa.gov/aboutepa/oswer.html*
  - Office of Wastewater Management *http://www.epa.gov/owm/*
  - Office of Water *http://www.epa.gov/ow/*
- US Geological Survey (water) *http://water.usgs.gov/*
- Centers for Disease Control and Prevention *http://www.cdc.gov/*
- American Water Works Association *http://www.awwa.org/*
- Water Environment Federation *http://www.wef.org/*
- California Department of Public Health *http://www.cdph.ca.gov/programs/Pages/DWP.aspx*
- California Department of Water Resources *http://www.dwr.water.ca.gov/*
- California State Water Resources Control Board *http://www.swrcb.ca.gov/*

The regulatory universe is filled with a mind-numbing array of acronyms—probably way too many. Rather than kill trees printing them in this text, the following URL is the USEPA's Web site containing more acronyms than you'll ever need. Trust me on this one. There's also a handy glossary.

- http://www.epa.gov/glossary/

The state of California also maintains a good acronym site at:

- http://www.dtsc.ca.gov/InformationResources/acronyms.cfm

REFERENCES

American Water Works Association. 2007. Public Involvement and Customer Communication. Statement of Policy on Public Water Supply Matters. Accessed October 2010 at http://www.awwa.org/files/about/OandC/PolicyStatements/2009PublicInvolvement.pdf. American Water Works Association, Denver, Colo.

American Water Works Association. 2010. Climate Change. Statement of Policy on Public Water Supply Matters. Accessed October 2010 at http://www.awwa.org/files/about/OandC/PolicyStatements/2010ClimateChange_1280772331328.pdf. American Water Works Association, Denver, Colo.

California Department of Water Resources. 2003. *Guidebook for Implementation of Senate Bill 610 and Senate Bill 221 of 2001 to Assist Water Suppliers, Cities, and Counties in Integrating Water and Land Use Planning*. Accessed October 2010 at http://www.water.ca.gov/pubs/use/sb_610_sb_221_guidebook/guidebook.pdf.

Crews, C.W. 2009. *Ten Thousand Commandments 2009* (http://cei.org/issue-analysis/2009/05/28/ten-thousand-commandments). Competitive Enterprise Institute, Washington, D.C.

Detels, R., R. Beaglehole, M.A. Lansang, and M. Gulliford, eds. 2002. *Oxford Textbook of Public Health*. Oxford University Press, Oxford, U.K.

EnvironmentalPrograms.net. 2010. Environmental Advocacy: Academic Requirements, Professional Outlook. Accessed October 2010 at http://www.environmentalprograms.net/guidance/env-advocacy/.

Environmental Working Group (EWG). 2010a. Over 300 Pollutants in U.S. Tap Water. Accessed October 2010 at http://www.ewg.org/ and http://www.ewg.org/tap-water/home.

EWG. 2010b. About the Environmental Working Group. Accessed October 2010 at http://www.ewg.org/about.

Fischhoff, B., S. Lichtenstein, and S.L. Derby. 1984. *Acceptable Risk*. Cambridge University Press, Cambridge, U.K.

Gallup Inc. 2009. *Approval of U.S. Congress Falls to 21%, Driven by Democrats*. Accessed October 2010 at http://www.gallup.com/poll/123491/Approval-Congress-Falls-21-Driven-Democrats.aspx.

Gallup Inc. 2010. *Congress Only Growing Less Popular With Americans*. Accessed October 2010 at http://www.gallup.com/poll/143054/Congress-Growing-Less-Popular-Americans.aspx.

HDR Engineering. 2001. *Handbook of Public Water Systems*. 2nd ed. John Wiley & Sons Inc., New York, N.Y.

Humber, W. 1879. *A Comprehensive Treatise of the Water Supply of Cities and Towns*. George H. Frost, Chicago.

Intergovernmental Panel on Climate Change (IPCC). 2010. Press Information: Opening Statement by Dr Rajendra Pachauri, Chairman of the Intergovernmental Panel on Climate Change at a press conference at the United Nations in New York. Accessed October 2010 at http://www.ipcc.ch/press_information/press_information.htm.

Johnson, S. 2006. *The Ghost Map*. Riverhead Books, New York.

Lindzen, R.S. 2009. Resisting Climate Hysteria. *Quadrant Online* G60. Accessed September 2010 at http://www.quadrant.org.au/blogs/doomed-planet/2009/07/resisting-climate-hysteria.

McGuire, M.J. 2006. Eight Revolutions in the History of US Drinking Water Disinfection. *Journal AWWA* 98:3, American Water Works Association, Denver, Colo.

Metcalf & Eddy. 2007. *Water Reuse*. Maidenhead: McGraw-Hill.

Michaels, P.J., and P.C. Knappenberger. 2009. Scientific Shortcomings in the EPA's Endangerment Finding from Greenhouse Gases. *Cato Journal* 29(3). Accessed October 2010 at http://www.cato.org/pubs/journal/cj29n3/cj29n3-8.pdf.

Minnesota Center for Environmental Advocacy. 2010. [Home page.] Accessed October 2010 at http://www.mncenter.org/.

Morrison, R. 2009. On 40th Anniversary of Cuyahoga Fire, Myths Loom Larger Than Reality. Accessed October 2010 at http://cei.org/news-release/2009/06/22/40th-anniversary-cuyahoga-fire-myths-loom-larger-reality). Competitive Enterprise Institute, Washington, D.C.

National Research Council. 1983. *Risk Assessment in the Federal Government: Managing the Process*, National Academy Press, Washington, D.C.

Rechtschaffen, C.L., and D.L. Markell. 2002. Reinventing the State/Federal Relationship in Environmental Enforcement. Accessed September 2010 at http://ssrn.com/abstract=327002. The Environmental Law Institute, Washington, D.C.

Roberson, A. 2009. Regulatory Update (10/27/09). American Water Works Association. Accessed October 2010 at http://www.awwa.org.

Scott, M. 2009. Cuyahoga River Fire 40 Years Ago Ignited an Ongoing Cleanup Campaign, The Cleveland Plain Dealer as reported on Cleve land.com. Accessed October 2010 at http://www.cleveland.com/science/index.ssf/2009/06/cuyahoga_river_fire_40_years_a.html.

Sheikh, B. 2010. *White Paper on Graywater*. American Water Works Association, Water Environment Federation, WateReuse Association, Alexandria, Va. Accessed September 2010 at http://www.awwa.org/files/Resources/Waterwiser/references/PDFs/GraywaterFinal%20Report2010.pdf. WateReuse Association, Alexandria, Va.

Snyder, S. 2008. Statement of Dr. Shane Snyder, Southern Nevada Water Authority before the Senate Subcommittee on Transportation Safety, Infrastructure Security, and Water Quality on Pharmaceuticals in the Nation's Water: Assessing Potential Risks and Actions to Address the Issue. Accessed October 2010 at http://www.awwa.org/files/GovtPublicAffairs/AWWA2008FlyinTestimonyPharmaceuticals.pdf.

Utah Division of Water Resources. 2010. Utah Water Jargon. Accessed September 2010 at http://www.ulct.org/ulct/docs/09June15-Water-Jargon.pdf.

USEPA (US Environmental Protection Agency). 1999. *25 Years of the Safe Drinking Water Act: History and Trends*. EPA 816R99007. Accessed October 2010 at http://www.epa.gov/history/topics/sdwa/index.htm.

USEPA. 2003. Public Involvement Policy of the US Environmental Protection Agency. Accessed September 2010 at http://www.epa.gov/publicinvolvement/pdf/policy2003.pdf. USEPA, Washington, D.C.

USEPA. 2004. *Guidelines for Water Reuse*. EPA/625/R-04/108. Accessed October 2010 at http://www.epa.gov/nrmrl/pubs/625r04108/625r04108.pdf.

USEPA. 2009a. Endangerment and Cause or Contribute Findings for Greenhouse Gases under Section 202(a) of the Clean Air Act. EPA

Docket EPA-HQ-OAR-2009-0171. Accessed October 2010 at http://www.epa.gov/climatechange/endangerment.html.

USEPA. 2009b. EPA's Endangerment Finding: Climate Change Facts. Accessed October 2010 at http://www.epa.gov/climatechange/endangerment/downloads/EndangermentFinding_ClimateChangeFacts.pdf.

USEPA. 2009c. News Releases—Air. EPA: Greenhouse Gases Threaten Public Health and the Environment / Science Overwhelmingly Shows Greenhouse Gas Concentrations at Unprecedented Levels Due to Human Activity. Release date: 12/07/2009. Accessed October 2010 at http://yosemite.epa.gov/opa/admpress.nsf/6424ac1caa800aab85257359003f5337/08d11a451131bca585257685005bf252!OpenDocument.

US Government Accounting Office (GAO). 1995. *EPA and the States: Environmental Challenges Require a Better Working Relationship.* US Government Accounting Office, Washington, D.C.

Wagner, T.P. 1999. *The Complete Guide to Hazardous Waste Regulations: RCRA, TSCA, HSWA, EPCRA, and Superfund.* 3rd ed. John Wiley & Sons, Inc. New York, N.Y.

Webster's Ninth New Collegiate Dictionary. 1986. Merriam-Webster, Springfield, Mass.

Wisman, Phil. 1985. EPA History (1970–1985). Prepared by the EPA Office of Public Awareness on the occasion of EPA's 15th Anniversary. Accessed October 2010 at http://www.epa.gov/history/topics/epa/15b.htm.

ADDITIONAL READING

American Public Health Association/American Water Works Association. 1925. *Standard Methods of Water Analysis*. 6th ed. American Public Health Association, New York, N.Y.

American Water Works Association. 2008. *The Business of Water: A Concise Overview of Challenges and Opportunities in the Water Market*. American Water Works Association, Denver, Colo.

Elsevier Science Inc. 1998. *Air Pollution Consultant Quick Reference Guide*. Elsevier Science Inc., Denver, Colo.

Hodges, L. 1977. *Environmental Pollution*. 2nd ed. Holt, Rinehart & Winston, New York, N.Y.

Metcalf & Eddy/AECOM. 2007. *Water Reuse*. McGraw-Hill, New York, N.Y.

USEPA. 2004. *Guidelines for Water Reuse*. USEPA, Washington, D.C.

Water Research Foundation (formerly Awwa Research Foundation). 2008. *Toxicological Relevance of EDCs and Pharmaceuticals in Drinking Water*. Water Research Foundation, Denver, Colo.

Water Research Foundation. 2004. *Risk Communication for Emerging Contaminants*. Water Research Foundation, Denver, Colo.

Water Research Foundation. 2008a. *Communicating the Value of Water: An Introductory Guide for Water Utilities.* Water Research Foundation, Denver, Colo.

Water Research Foundation. 2008b. *State of Knowledge of Endocrine Disruptors and Pharmaceuticals in Drinking Water.* Water Research Foundation, Denver, Colo.

INDEX

25 Years of the Safe Drinking Water Act: History and Trends, 38
303(d) Program, 47–48, 49
404 permit, 43
1986 SDWA amendments, 34–35
1996 SDWA amendments, 35–38

A

AB 32 (Global Warming Solutions Act of 2006), 79–80
ACM (asbestos-containing materials), 66
activists, 19–20
adenovirus, 128
advocacy, environmental, 113–115
AFOs (animal feeding operations), 44
agricultural customers, 18
AGW (anthropogenic global warming), 79–82
aircraft drinking water, 39–40
air quality standards, 57
American Clean Energy and Security Act of 2009, 80
American Water Works Association (AWWA), 74, 82, 106–107, 132
animal feeding operations (AFOs), 44
anthropogenic global warming (AGW), 79–82
Aqua America, Inc., 102
Arkansas River Coalition (ARC), 52
Arkansas River Compact, 100
asbestos-containing materials (ACM), 66
Audubon, James J., 26
AWWA (American Water Works Association), 82, 106–107

B

beneficial use, 67–68
Bioterrorism Preparedness and Response Act of 2002, 86
bottled water, 38–39
Broad Street Pump, London, 31–32
Brownfields Program, 91
"bugs and bunnies", 76–79

Burroughs, John, 26
business customers, 18

C

CAA (Clean Air Act), 55–60
  future regulations, 126
  overview, 55–59
  Risk Management Plan Rule, 59
  utilities and, 59–60
CAFOs (concentrated animal feeding operations), 44
caliciviruses, 128, 132
California
  California Air Resources Board (CARB), 60, 79
  Global Warming Solutions Act, 79–80
  graywater standards, 74
  law codes, 7
  recycled water in, 68–69
  Senate Bills 610/221, 83–85
  water law in, 97–98
  water supply, 77–79
  Western Municipal Water District, 34–35, 102, 103
California Code of Regulations (CCR), 7
Campylobacter jejuni, 128
Canada, water agreements with, 100
Carson, Rachel, 26
CCL (Contaminant Candidate List), 127–129
CCR (California Code of Regulations), 7
CCR (Consumer Confidence Report), 108
CERCLA (Comprehensive Environmental Response, Compensation, and Liability Act), 64, 66, 88, 91
CFR (Code of Federal Regulations), 7
chemical contaminants, 127–128
chlorine
  DBPs, 37
  RMP Rule, 59
cholera, 31
classical smog, 55
Clean Air Act (CAA), 55–60
  future regulations, 126
  overview, 55–59
  Risk Management Plan Rule, 59
  utilities and, 59–60
Clean Water Act (CWA), 41–53
  303(d) program, 47–48
  future regulations, 126
  MS4 program, 50–51
  national pretreatment program, 46–47
  nonpoint sources, 51–52
  NPDES, 43–46
  overview, 41–43
  SDWA and, 53
  stormwater regulations, 48–51
  TMDLs, 47–48
  water quality standards, 47–48
climate change, 82
CNMPs (Comprehensive Nutrient Management Plans), 44, 46
Code of Federal Regulations (CFR), 7
Colorado
  Colorado River Compact, 98–99
  HB 1141, 85–86
  water law in, 97
comment period, public, 9
communications, public
  emergency notifications, 108–110
  overview, 107–110
  risk communication, 111–113
  routine communication, 108
"Community Right-to-Know" reporting, 65
compacts, water, 98–100
compliance, programmatic costs of, 11
Comprehensive Environmental Response, Compensation, and Liability Act (CERCLA), 64, 66, 91
Comprehensive Nutrient Management Plans (CNMPs), 44, 46
concentrated animal feeding operations (CAFOs), 44
construction grants program, 43
construction-site stormwater discharges, 50
Consumer Confidence Report (CCR), 108
Contaminant Candidate List (CCL), 127–129
contaminants
  1962 US Public Health Service Standards, 33

CCL, 127–129
chemical, 127–128
EDCs, 130–132
emerging contaminants, 129
microbial, 128–129
NDMA, 129–133
pathogens, 132–133
perchlorate, 132
USEPA standards for, 31
Vanishing Zero concept, 120–123
"cooperative federalism", 88
cost–benefit analysis, 11
costs, regulatory, 10–11
Crapo, Mike, 126
"crying Indian" commercial, 41–42
cryptosporidiosis, 35–36
Cryptosporidium parvum, 35–36
custody protocols, 119–120
customers
agricultural, 18
business, 18
general public, 17–18
industrial, 18
Cuyahoga River fire, 26
CWA (Clean Water Act), 41–53
303(d) program, 47–48
future regulations, 126
MS4 program, 50–51
national pretreatment program, 46–47
nonpoint sources, 51–52
NPDES, 43–46
overview, 41–43
SDWA and, 53
stormwater regulations, 48–51
TMDLs, 47–48
water quality standards, 47–48

D

DDT (dichloro-diphenyl trichloroethane), 26
delta smelt, 77–78
disinfection by-products (DBPs), 37
draft publications, 9
"Drinking Water Treatment Effluent Guidelines", 53
drought, 77–78

E

EDCs (Endocrine Disrupting Compounds), 130–132
EIS (environmental impact statement), 28
elected officials, 14–16
emergency notifications, 108–110
Emergency Planning and Community Right-to-Know Act of 1986 (EPCRA), 65
emerging constituents, 130
emerging contaminants, 129
Endangered Species Act (ESA), 76–79
Endangerment Finding (2009), 82
Endocrine Disrupting Compounds (EDCs), 130–132
enforcement, state regulatory, 89–91
enterovirus, 128
EnvCAP Web site, 92–93
environmental advocacy, 113–115
environmental drought, 77
environmental impact statement (EIS), 28
environmental laboratories. *See* laboratories
environmental policy, 9–10
Environmental Working Group (EWG), 114–115
EPA and the States: Environmental Challenges Require a Better Working Relationship, 89
EPCRA (Emergency Planning and Community Right-to-Know Act of 1986), 65
ESA (Endangered Species Act), 76–79
E.S. Babcock & Sons Laboratories, Riverside, California, 118–120
Escherichia coli, 128, 133
ethics, laboratory code of, 119–120
EWG (Environmental Working Group), 114–115

F

Federal Register, 1–2
federal/state regulatory relationship, 87–91
Federal Water Pollution Control Act (1948), 41

filtration, 31, 32
"first in time, first in right", 97–98
Fischhoff, B., 112–113
flammable chemicals, 59
future regulatory concerns, 125–133
  EDCs, 130–132
  emerging contaminants, 129
  NDMA, 129–133
  pathogens, 132–133
  perchlorate, 132

G

Gallup poll, September 2010, 15–16
general public, 17–18
germ theory of disease, 32
The Ghost Map (Johnson), 31–32
Giardia lamblia, 36
global warming, 79–82
  AWWA and climate change, 82
  complications, 81–82
  overview, 79–81
Global Warming Solutions Act of 2006 (AB 32), 79–80
graywater, 73–74
Great Lakes Water Quality Agreement, 100
green maze, 5–12
  defined, 5–6
  regulatory structure, 6–11
    from laws to regulations, 7–9
    permits, 9–10
    policy, 10
    regulatory costs, 10–11
groundwater replenishment, 72
Guidelines for Water Reuse (USEPA 2004), 70

H

HAPs (hazardous air pollutants), 58
Hardinge filtration plant, California, 34, 35
Harmon Industries, Inc. v. Browner, 90–91
hazardous air pollutants (HAPs), 58
hazardous materials, 64–65
hazardous waste
  categories of, 62
  defined, 61–62
hazardous waste regulations, 61–66
  hazardous materials management, 64–66
  history of and RCRA, 63–64
  right-to-know regulations, 64–66
  Superfund, 64
  water/wastewater utilities and, 66
Helicobacter pylori, 128
Hepatitis A virus, 128
history of environmental regulations, 23–28
  1960s and 1970s, 26–28
  before modern environmental movement, 26
  overview, 23–25
House Bill 1141, Colorado, 85–86
hybrid systems of water law, 97–98

I

ICR (Information Collection Rule), 37
Illinois, law codes in, 7
income, relationship to regulations, 27, 28
industrial customers, 18
industries
  discharge regulations under CWA, 42
  pretreatment programs, 46–47
Information Collection Rule (ICR), 37
international water compacts, 100
Internet, 19
Internet resources, 135–136
interstate water compacts, 98–100
investor-owned utilities (IOUs), 102
irrigation, 69, 71

K

Knappenberger, Paul C., 82

L

laboratories, 117–123
  overview, 117–119
  protocols, 119–120
  Vanishing Zero, 120–123
laws, passing of, 7–9
Legionella pneumophila, 129, 133
Lindzen, Richard S., 80–81
Local Emergency Planning Committees (LEPCs), 65
London's Broad Street Pump, 31–32
longfin smelt, 78

M

Markell, David L., 88
mass media, motivation of, 18–19
maximum achievable control technology (MACT), 58
maximum contaminant level goal (MCLG), 31
M/DBP Rules (Microbial/Disinfection By-Product Rules), 38
media, 18–19
Mexico, water agreements with, 100
Michaels, Patrick J., 82
microbial contaminants, 128–129
Microbial/Disinfection By-Product Rules (M/DBP Rules), 38
Microsoft Word templates for public notices, 109–110
MS4 (Municipal Separate Storm Sewer Systems) program, 50–51
Muir, John, 26
Municipal Separate Storm Sewer Systems (MS4) program, 50–51
Mycobacterium avium, 129

N

Naegleria fowleri, 129, 133
National Academy of Sciences contaminant study, 33
National Ambient Air Quality Standards (NAAQS), 57
National Emission Standards for Hazardous Air Pollutants (NESHAPs), 58
National Environmental Policy Act (NEPA), 28, 76
National Pollutant Discharge Elimination System (NPDES), 42, 43–46, 49–50, 117–118
National Pretreatment Program, 42
NDMA (N-nitrosodimethylamine), 129–133
new source review (NSR), 58–59
nonattainment areas, 57
nonpoint sources, 51–52

O

Office of Solid Waste and Emergency Response (OSWER), 91
officials, elected, 14–16
ordinances, 101–104
overfiling, 90–91

P

Pachauri, Rajendra, Dr., 81
para-typhoid fever, 29–30
participation, public, 105–107
Passaic Valley Water Commission (New Jersey), 118
Pasteur, Louis, 32
pathogens, 132–133
pea-soup smog, 55
perchlorate, 132
permits, 9–10
personal income, relationship to regulations, 27, 28
pharmaceuticals and personal care products (PPCPs), 130
photochemical smog, 55
Pinchot, Gifford, 26
planned indirect potable reuse, 72
policy, 10
pollutants in stormwater, 52
PPCPs (pharmaceuticals and personal care products), 130
press, motivation of, 18–19
pretreatment programs, 46–47
primacy, 30–31
primary standards, 57
prior appropriation, 97–98
programmatic costs of compliance, 11
promulgation, 8–9
protocols, laboratory, 119–120
public, environmental regulations and, 105–115
    environmental advocacy, 113–115
    public communication, 107–110
    public participation, 105–107
    risk communication, 111–113
public comment period, 9
public communication
    emergency notifications, 108–110
    overview, 107–110
    routine communication, 108
public health, 107
Public Health Service, US, 32–33
Public Involvement and Customer Communication Policy Statement (AWWA), 106–107

Public Involvement Policy (USEPA), 105–106
public notification, 108–110
public participation, 105–107

R

Rapanos et ux., et al. v. United States, 126
RCRA, 63, 66, 75, 88, 90
Rechtschaffen, Clifford L., 88
recycled water, 67–74
  federal regulations, 70–71
  graywater, 73–74
  indirect potable reuse, 71–72
  overview, 67–68
  process, 68–69
  quality of, 70–71
  uses of, 69
regulations, 75–86. *See also* future regulatory concerns; public, environmental regulations and; state environmental regulations
  Bioterrorism Preparedness and Response Act of 2002, 86
  Endangered Species Act, 76–79
  global warming, 79–82
  history of, 23–28
  making of, 7–9
  overview, 75
  relationship to personal income, 27, 28
  society of, 1–2
  timeline of, 24–25
  water supply availability requirements, 83–86
regulators, 16–17
regulatory costs, 10–11
regulatory drought, 78
regulatory process, 8
regulatory promulgation activities, 8–9
regulatory structure, 6–11
  from laws to regulations, 7–9
  permits, 9–10
  policy, 10
  regulatory costs, 10–11
reservoir augmentation, 72
Resource Conservation and Recovery Act (RCRA), 63, 66, 75, 90
Revised Public Notification Handbook (USEPA), 109
riparian rights, 96–97
risk assessment, 111
risk communication, 111–113
risk management, 111
Risk Management Plan Rule (RMP Rule), 59
routine communication, 108
Ruckelshaus, William, 20–21

S

Safe Drinking Water Act (SDWA), 29–40
  1986 amendments, 34–35
  1996 amendments, 35–38
  coverage of, 38–40
  and CWA, 53
  future regulations, 126–127
  overview, 29–31, 32–33
  preceding factors, 30–32
  public communications requirements of, 107–110
  USEPA and history of, 38
Salmonella enterica, 129
SARA (Superfund Amendments and Reauthorization Act) Title III, 65
SDWA. *See* Safe Drinking Water Act (SDWA)
Secondary Drinking Water Regulations, 31
secondary standards, 57
Senate Bill 221, California, 83–85
Senate Bill 610, California, 83–85
September 2010 Gallup poll, 15–16
SERCs (State Emergency Response Commissions), 65
Shigella sonnei, 129
Silent Spring (Carson), 26
smog, 55
snail darter, 77
Snow, John, Dr., 31, 32
Snyder, Shane, 122–123, 131–132
Solid Waste Agency of Northern Cook County v. United States Army Corps of Engineers et al, 126
source control, 46
special districts, 102
stakeholders, 13–22
  activists, 19–20
  customers
    agricultural, 18

business, 18
general public, 17–18
industrial, 18
elected officials, 14–16
media, 18–19
overview, 13–14
regulators, 16–17
USEPA, 20–22
utilities, 17
State Emergency Response Commissions (SERCs), 65
state environmental regulations, 87–93
enforcement of, 89–91
federal/state regulatory relationship, 87–91
state regulatory structures, 91–93
Superfund law, 91
state regulatory structures, 91–93
State Water Project, California, 77–78
Stormwater Pollution Prevention Plan (SWPPP), 44
stormwater regulations, 48–51
Superfund, 65, 91
Surface Water Treatment Rule, 34–35
SW-846, 63–64
SWPPP (Stormwater Pollution Prevention Plan), 44

T

Tellico Dam, TVA's, 77
templates for public notices, 109–110
Tennessee Valley Authority's (TVA) Tellico Dam, 77
tertiary-treated water, 69
"the purple book", 69
THMs (trihalomethanes), 37
Tier 1 violations, 109
Tier 2 violations, 109
Tier 3 violations, 109
timeline of regulations, 24–25
Title IV permits, 58
Title V permits, 59
Total Coliform Rule, 34
total maximum daily load (TMDL), 41–42, 47–48, 49
toxic chemicals, 59
Toxic Substances Control Act (TSCA), 66
transite pipe, 66
trihalomethanes (THMs), 37
TVA v. Hill, 77
typhoid fever, 29–30

U

"underground law", 10
Unified Agenda (UA), 126
Union Carbide plant, India, 64
United States v. Power Engineering Co., 91
unplanned indirect potable reuse, 72
Unregulated Contaminant Monitoring Rule (UCMR), 130
U.S. Environmental Protection Agency (USEPA)
control of nonpoint sources, 52
Federal Register entry for, 2
and history of SDWA, 38
Microsoft Word templates for public notices, 109–110
overfiling, 90–91
overview, 20–22
Public Involvement Policy, 105–106
Revised Public Notification Handbook, 109
and state regulatory enforcement, 89–91
state-specific regulatory Web site, 92–93
US Public Health Service, 32–33
utilities
CAA and, 59–60
IOUs, 102
ordinances, 101–104
purpose of, 107
relationship of regulator with, 16–17
RMP Rule, 59
as stakeholders, 17

V

Vanishing Zero, 120–123
vulnerability assessments (VAs), 86

W

wastewater utilities
CAA and, 59–60
motivation of, 17
ordinances, 101–104
purpose of, 107
relationship of regulator with, 16–17

wastewater utilities (*continued*)
RMP Rule, 59
as stakeholders, 17
water compacts, 98–100
water law, 95–100
hybrid systems of, 97–98
international water compacts, 100
interstate water compacts, 98–100
overview, 95–96
prior appropriation, 97–98
riparian rights, 96–97
water quality standards, 47–48, 49
water reclamation. *See* recycled water
water reuse. *See* recycled water
water supply
availability requirements, 83–86
California SB 610/SB 221, 83–85
Colorado HB 1141, 85–86
overview, 83
of California, 77–79
Water Supply Assessment (WSA), 84–85
water utilities
CAA and, 59–60
ordinances, 101–104
purpose of, 107
relationship of regulator with, 16–17
RMP Rule, 59
as stakeholders, 17
Western Municipal Water District, California, 34–35, 102, 103
Wilson, Alexander, 26
WSA (Water Supply Assessment), 84–85

www.ingramcontent.com/pod-product-compliance
Lightning Source LLC
Chambersburg PA
CBHW081126180926
51611CB00004B/62
* 9 7 8 1 5 8 3 2 1 8 1 5 0 *